MARTIN FIELD

# CREATING COMMUNITY-LED AND SELF-BUILD HOMES

A Guide to Collaborative Practice in the UK

POLICY PRESS SHORTS POLICY & PRACTICE

First published in Great Britain in 2020 by

Policy Press
University of Bristol
1-9 Old Park Hill
Bristol
BS2 8BB
UK
t: +44 (0)117 954 5940
pp-info@bristol.ac.uk
www.policypress.co.uk



British Library Cataloguing in Publication Data
A catalogue record for this book is available from the British Library

ISBN 978-1-4473-4439-1 paperback
ISBN 978-1-4473-4441-4 ePub
ISBN 978-1-4473-4440-7 ePdf



Cover design by Policy Press
Front cover image: iStock

Printed and bound by CPI Group (UK) Ltd, Croydon, CR0 4YY

# Contents

# Figure and tables

## Figure



## Tables

# Acknowledgements

This 'Policy and Practice' publication explores the varieties and diversity in how local people have been engaging with housing and neighbourhood initiatives across the UK.

The source material has principally been gained from engagement and collaboration with a substantial range of local and community-centred practices over many years. A huge debt is acknowledged to them all and to the ongoing work of the regional and national bodies within the community-led housing sector. If the details provided on different forms of collaborative housing include some lengthy lists, this is deliberately to reflect on the richness now found within this sector that is too easily overlooked when only a brief example or two are given attention.

All website information was operational at the time the work was published, however flows of information are always dynamic and links will need to be checked accordingly. All tables and figures are from the author's own work, except Figure 3.1, used under common licence.

My thanks are due to Policy Press for their patience in looking through the different versions of this work. The overall ideas started out as something far more extensive and ambitious but have been better served by the style required in the 'Short' series.

And a special thank you to Dinah and to John for their faith in me.

# Introduction: people providing homes for themselves in the UK

This is a book about people in the UK taking charge of their own homes. At a time when there is so much dissatisfaction with the kinds of homes (and their condition) available for sale or rent in the UK, the publication sets out to explain the ways people have wished to house themselves, in places of their own choosing and their own designs.

The following chapters provide a framework to analyse what people have been doing in a wide variety of activities – as individuals, in groups, or on behalf of their local communities – to create homes that can satisfy many kinds of ambition:

Eco-housing COHOUSING COMMUNES
Co-operatives SELF-BUILD Alternative
Community Land Trusts INTENTIONAL COMMUNITIES Travellers
New Age Self-Help Housing Green Architecture
CUSTOM-BUILD Ecovillage Tenant Management
SELF-PROVIDED Group-build COMMUNITY-LED
Low Impact Development Trusts Live-aboards
Intentional Neighbourhoods SELF-COMMISSIONING
Short-Life collaborative housing The Commons
ALMOs Community Gateways Stewardship
MUTUALS Squatting BARGE DWELLERS

maf.2017

Source: © M. Field 2019

The UK's housing sector is increasingly full of uncertainties regarding how people can find and maintain a satisfactory home in which they can thrive. Headline reports routinely note the diminishing control many households have on their domestic circumstances – fewer and fewer households in owner-occupation; rising numbers forced to live in accommodation provided by an uncompromising private rented sector; and prolonged time spent by young adults living in the parental home.

There are also competing debates at national and international levels on the ends to which modern housing provision could or should be tailored – whether its primary function remains the asset basis for a welfare provision that can deliver affordable housing to households unable to find other suitable accommodation, or if it should be focused upon real estate investment to stimulate wider property-based wealth within local communities.

A brief summary of how the UK's housing sector plans for the construction of homes and delivery of neighbourhood services would include the following observations:

- The UK's present culture of housing provision is a tried and tested framework for mainstream participants to plan and provide housing and neighbourhood services.
- It is intrinsic to the nature of the UK's 'open market' practices that, notwithstanding some occasional innovation, existing relationships remain unchanged until dominant interests accept reasons to change.
- Ideas for new forms of, or gaps in, housing provision are best assessed and addressed by the experienced partnerships and frameworks already in place – the extent of any local demand (such as represented by community-led housing solutions) is not proven, or will only be present on a very modest scale.
- The established networks which interact on land, finance and construction services are not ones in which 'community'

> projects can easily compete – proposals for solutions from the grassroots of local communities are not seen to have the governance skills or practical expertise to drive new projects forward.

This publication centres around the nature and ways in which the individual and collaborative actions of people in the UK have sought to challenge the above positions. An underlying theme is an awareness of the tension between how local people can obtain lasting and secure homes and neighbourhoods at *affordable* costs and how there should be *accountability* to local communities and their local residents for the homes and places created. What follows are extensive descriptions of the innovative responses of local people to achieve the homes and neighbourhoods that can meet their personal and collective ambitions, plus ideas on the kinds of benchmarks for how the public, private and community sectors might be accommodating of such activities.

# ONE

# Identifying motivation at the grassroots

## The historical context

Whilst the main purpose of this publication is admittedly to detail and explore *contemporary* settings for how people in the UK are shaping their homes and neighbourhoods, it is important to recall the rich legacy that new projects might use for inspiration. Local communities have consistently adapted to key circumstances of the day in order that the benefits from using local resources can be properly shared – whether that be the appropriation of land by the Diggers and Levellers of the 17th century or the 'community buyouts' of Scottish highland estates in the 21st.

The persistent nature of how people have sought to create their own homes can always be understood as a response to key and prevailing issues of the time. An early 19th century focus of workers within industrial trades challenging their exclusion from prevailing political and financial systems resulted in the creation of the first 'friendly societies', based on a simple premise that if a group of people contributed to a mutual fund, they could then receive benefits at a time of future need.[1] Working class families without access to the main banking system of the age established the first building societies from 1775, for members to pool funds for purchasing land and building houses. Over 250 societies had been created by 1825, originally set up as *terminating* societies, which closed when all the members had been housed or had purchased land for that purpose. In the 1830s and 1840s the *permanent* building society gradually emerged, continually taking in new members as earlier ones completed their purchases and became suitably housed.

The middle years of the 19th century also witnessed the emergence of Chartism as an aspirational working class movement for political reform. It was named after the People's Charter of 1838 and aimed to reform the democratic nature of national and local politics. In 1843 the Chartist Co-operative Land Company was established (later called the National Land Company) to enable working people to acquire land for their own housing and to challenge the private appropriation of land by Enclosure Acts. Funds collected from contributions by workers were used to purchase rural sites, which were then subdivided into smaller plots for occupation by households chosen through drawing lots. Between 1844 and 1848, five estates were purchased, subdivided and cottages were built: former Land Company houses remain in use in Oxfordshire, Worcestershire and Gloucestershire.

Parallel to these endeavours for new settlements of individually owned properties, the Rochdale Pioneers were meeting their members' housing needs through other innovative collaborations. This led to the first co-operative housing being built in 1861 by the Rochdale Pioneer Land and Building Company, which went on to provide 84 homes for its members by 1867.

Gillian Darley's work *Villages of Vision* provides evocative details on the many kinds and scales of community-centred settlements that emerged around the nation in the later years of the 19th century – not least the well-publicised 'model village' settlements like New Lanark, Saltaire, Port Sunlight and Bournville. The superior living standards available within such planned settlements certainly represented improvements in domestic circumstances; however, such developments were firmly controlled by the representatives of their respective founding figures. There was limited opportunity for tenants or residents to shape what dwellings were being offered, and on what terms. Other kinds of communities therefore emerged based upon different political ideals than the rather authoritarian kind of governance found in the model villages. These were interested in a more communal and co-responsible life, with a sense of how residents could live in local places through holding shared values in common with each other. Denis Hardy and Lynne Pearson have listed the diverse communities of 'utopian' socialists, anarchists, Tolstoyans, communists and others. Some are still evident: the Whiteway Colony in Gloucestershire continues to exist over 100 years after being originally set up as an 'anarchic settlement' with explicit principles against any individual ownership of land. It is (in)famous for holding a ceremonial public burning of all the title deeds to the estate in order to destroy any legal or future basis for individuals to claim ownership of its land.

The focus on acquiring land for the large-scale benefit of future local communities was a fundamental driver behind the creation of the first garden cities, such as Letchworth Garden City, in the period at the start of the 20th century and immediately prior to the First World War, and a subsequent creation of administrative structures to extend this drive to new 'garden suburbs'. These ideals underpinned intentions to safeguard the community stewardship of local land and other facilities and informed the kind of co-operative housing developments built by tenant co-partnerships, starting in 1901 with the founding of the first co-partnership co-operative at Bentham Garden Suburb in Ealing, and subsequently other examples, such as at Hampstead Garden Suburb in Hampstead.

The end of the First World War in 1918 brought a huge demand for new housing and community growth to the fore, with the campaign slogan of 'Homes for Heroes' repeated throughout Britain, in part due to the recognition of the substandard slum properties still evident across much of the nation. It was at this time that the first local council housing provisions emerged and the first suburban estates built from public funds were constructed, yet there remained a strong individualistic attraction in the idea of possessing a few square yards on which to build something more permanent – small holiday homes, country retreats or even a cherished 'smallholding'.

In sparsely populated places, especially those close to more isolated coastal areas, plotland developments began to appear, the term 'plotland' being coined for those places where land was divided and sold in small plots (often in unorthodox ways), not infrequently from an enforced sale of bankrupt farms at knockdown prices. These areas frequently developed characteristics as local arcadias of landscaped enclaves and grassy tracks, randomly filled with a collection of dwellings made from army huts, old railway coaches, sheds, shanties and diverse chalets,[2] powerfully recorded by Denis Hardy and Colin Ward in their work *Arcadia for All*.

Immediately after the Second World War, with a pressing need for accommodation of any kind, an increasing number of families became involved in large-scale squatting campaigns, particularly to make use of decommissioned military camps and other decommissioned dwellings. There was an acute shortage of properties for households leaving service life, and for those whose homes had been destroyed by bombing. Local ad hoc 'committees', often of ex-service personnel, installed homeless families in vacant houses under cover of night, there to remain until the absentee owners could initiate legal proceedings against them. In the following years, such action grew to include the forced occupation of hundreds of army and air force camps no longer needed for military purposes. Settlers in the Sheffield area formed a Squatters' Protection Society and linked up with other pioneer squatters from similar urban areas like Scunthorpe. The government announced by October 1946 that 1,038 camps in England and Wales had been occupied by 39,535 people, and that 4,000 people had squatted in Scotland. Later in the year the Ministry of Works offered the Ministry of Health (then the department responsible for housing) an additional 850 former service camps to use.

Substantial numbers of permanent new dwellings were clearly essential, and the post-war government instigated an entire revised town planning regime as an integral element of what was sought from the new welfare state, with a programme for building new towns across the UK from the later 1940s onwards. In the new towns themselves, the ethos was substantially that housebuilding and neighbourhood development should be the responsibility of the local state and its local development corporations. Nevertheless, construction within the urban settlement areas during the 1950s and 1960s still saw contributions by private engagements through organised self-build groups where future residents pooled their trades and skills together for common benefit. The promotion of self-build plots in new towns was modest in scale, at times just targeted at attracting senior executive households into specific local areas – in later developments, like in Milton Keynes, more opportunities were included for self-build activity, as this had been receiving wider coverage in the public debate on housing opportunities at that time.

As the euphoria of the first new town developments began to wane and core political support began change, criticism of local authority competencies to manage and maintain the public sector housing stock became more strident. Opportunities were sought by council tenants to establish new tenant management organisations, set up to take over the direct management of local and neighbourhood housing services by local tenants themselves. Government assistance programmes in the 1970s and 1980s subsequently promoted grant finance towards the management of homes for low- and middle-income families from new housing co-operatives and other self-help housing organisations, at times set up as 'short life' co-operatives for temporary use of properties emptied under plans for wider urban redevelopment programmes or left under-used by absentee owners.

A further impetus to co-operative housing activity came in the 1960s with the promotion and development of co-ownership schemes – property built and managed by a collective group of residents, paying a monthly rental to redeem the mortgage originally borrowed to build their homes. The affordability of the tenure was in turn assisted by government tax relief on the mortgage loan. Over 40,000 co-ownership homes were built and operated up to the early 1980s. From the 1990s onwards, when state support began to diminish, larger housing associations become a more favoured social housing delivery mechanism, rather than through small bodies like the co-ops. Some co-operatives subsequently developed other independent means to raise building funds, such as innovative ideas for new loanstock arrangements.

The focus on doing things in common with others was also apparent within the 1960s and 1970s by a rising interest to create 'alternatives' to orthodox or mainstream social settings, informed not least by interest in the international 'hippy' movement. A classic (and stereotypical) feature of this time was the rise of communes as a basis for alternative communities between people sharing similar ideas and values, often as a means to avoid any adoption of conventional forms of individual ownership or property speculation (even drawing from precedents like the Diggers protests in the Jacobean age). As memorably summarised by Chris Coates in his two works *Utopia Britannica* and *Communes Britannica*, what developed from the 1960s onwards was a variety of permanent and transient communities in which families and groups explored the latest ideas about personal and social relationships and the impact of these upon political commitments.

Whilst substantial amounts of public housing were still built into the 1970s, it was also a time of growing political and social engagement with ideas to support private sector home ownership, an interest that can be seen to have initiated the present-day dominance by private sector developers and housebuilding firms over new supply. An individualistic appeal of building a new property for one's own household grew in tandem with this, and independent self-build consultants found sites for small schemes on which groups of self-builders would collaborate and then purchase the properties on completion of the scheme. By the 1980s further initiatives for low-tech self-build (like the Segal simplified timber-frame technique) and community self-build schemes were set up in order to help households on lower incomes take part in such activity, with programmes of public grants being available to local groups and their development partners – usually local housing associations.

Community-led innovations continued to emerge during the 1980s and 1990s, and many local initiatives were progressed by new community development trusts that could function as the means to focus local concerns into activity that could bring commercial and financial benefits back to local communities. Concerns also emerged on how construction and development projects might reduce their impact upon the health of local people and upon the natural environment. Low-impact housing schemes promoted 'green' designs and eco-architecture, and proposals for large and holistic ecovillages emerged from groups of eco-minded professionals and prospective residents seeking to implement such ideals on significant scales. There was also the first successful examples of UK cohousing neighbourhoods, demonstrating the dynamics to be achieved from collaborative engagements between households at a slightly smaller scale.

Finally, now well into the 21st century, new initiatives continue to be explored, such as the community land trusts established by local communities seeking to secure affordable housing as a long-term local asset, or to be the vehicle for acquiring larger areas of land, such as the purchase of the island of Eigg by a community heritage body set up for that end.

## The contemporary context

Initial response to the 'financial crisis' in western industrial societies during 2007 and 2008 brought forward a number of observations from political commentators on the apparent failure of international housing and finance markets to meet social needs in a sustainable, efficient or prudent way. The first months of that crisis included frank public criticism of the malevolent aspect of prevailing 'market forces', and how the actions of influential speculators had proved to be so destructively self-serving. Yet in a short space of time, a willingness to explore what could remedy the causes of a 'failing market' has been subsumed within an alternative focus and narrative of a 'crisis' in providing enough new properties, especially affordable and accessible properties, for the future. Even the Government White Paper from 2017 – starkly titled 'Fixing Our Broken Housing Market' – quickly moves from a critical stance on housing provision being unduly reliant on the dominant interests of volume housebuilders to a set of proposals for increasing future supply largely through a greater level of output from the same providers. The conclusions of that White Paper were that an increased number of new homes would be sufficient to transform the failing market and usher in a more affordable future, rather than the market needing more fundamental change.

This focus on a perceived crisis in delivery levels obstructs a real challenge being promoted on what could persist as underlying factors of the original market failure. Evidence continues to come forward of the shortcomings in conventional planning and construction undertaken by the same suppliers that are still courted by government policy. Among these shortcomings are the significant limitations in housing design and the poor quality of new neighbourhoods,[3] where a lack of basic community facilities raises serious concern that such areas will in no way be 'sustainable' places for future social interactions. Nor does the 'crisis in delivery' narrative adequately account for how the UK's housing and property markets have emerged as even more expensive and unaffordable to many households in the years following the financial crash – as Bob Colenutt has recently highlighted, there is a real naivety in the simplistic belief that having more housing units available will by itself reduce market prices, when so much of the UK market context is controlled by the interests of the biggest housing suppliers. The Affordable Housing Commission[4] contains some excellent contemporary analyses of what 'affordable' may now mean to UK households and other stakeholders, but it has been quiet on macroeconomic considerations that could inform a critique of *why* there is such an apparent shortage of properly affordable dwellings and *how* to respond so that future households will not see housing costs consume such an unsustainable proportion of their finances.

It is pertinent to recall observations made more than a century ago on the history of inequalities in socioeconomic conditions between social classes in the UK and their respective access to political and economic power:

> The more closely any period of history is studied, the more clearly does it appear that the mistakes and troubles of an age are due to a false spirit, an unhappy fashion in thought or emotion, a tendency in the human mind to be overwhelmed by the phenomena of the time, and to accept those phenomena as the guide to conduct and judgement, instead of checking and criticising them by a reasoned standard of its own. [People] come to think that it is their business to explain, rather than to control the forces of the hour.[5]

One of the less well-publicised responses to the UK's financial crisis of 2007–8 is that a number of UK organisations representing housing and neighbourhood interests at the 'grassroots' level saw the opportunity to raise discussion beyond a narrative that would focus solely on a 'crisis' and promote a broader debate about the holistic nature and accountability of the UK's housing future.

An embryonic coalition was formed, informally termed the 'Mutual Housing Alliance',[6] to collaborate on ways to promote an increase in practical opportunities for future local housing projects. There was an acceptance by members in this Alliance that each could have a discrete focus on what initiatives local people might wish to instigate at local levels – it could be facilitating housing to be collectively owned or managed (such as by co-operatives or trusts), or a focus on the building of new properties by individuals or by neighbourhood groups (such as with self-build or cohousing groups), or even simply to renovate and reuse buildings in disrepair. Yet it was also acknowledged that a collaboration of work between bodies would provide a complementary and common view on how the core accountability of all kinds of successful local activities should rest with the households and communities seeking the local change.

Members of the Mutual Alliance subsequently contributed to a variety of government initiatives and working parties instigated by both Labour and Conservative governments, and a narrative began to emerge demonstrating to a wider audience what 'community-led' activity represents at its heart. This was subsequently enshrined in definitions commissioned by members of the Alliance from the Building and Social Housing Foundation (BSHF)[7] to confirm that 'community-led' housing is:

> [where] people are involved in meeting their own housing needs and wants. The route taken depends on things like the nature of the demand, available resources,

> location and type of activity. Approaches encompass new build, regeneration and the use of existing buildings. Community-led housing groups may form on the basis of a geographical connection or something else they have in common [....] the 'community-led housing' sector supports people to decide how they want to live, and make it happen.[8]

The coherence between the common interests of different forms of activities and organisations within the 'community-led housing sector' was also clarified in a working definition of what principles constitute the practical workings of such projects:

> The legal form and activities of each community-led housing scheme depend on the outcomes needed, but share common principles:
> a) the community is integrally involved throughout the process in key decisions like what is provided, where, and for who. They don't necessarily have to initiate the conversation, or build homes themselves;
> b) there is a presumption that the community group will take a long-term formal role in the ownership, stewardship or management of the homes;
> c) the benefits of the scheme to the local area and/or specified community group are clearly defined and legally protected in perpetuity.[9]

This understanding of what is intrinsic to local people's housing and neighbourhood projects has subsequently informed other extensive work and collaborations with government parties, undertaken by a secondary grouping of bodies that were initial members of the Mutual Alliance. From 2017 a fourfold practical collaboration between the National Community Land Trust Network, the UK Cohousing Network, the Confederation of Co-operative Housing and Locality has created the 'Community Led Homes' alliance, and helped steer the recent political and financial support from central government provided through the Community Housing Fund.[10] More comment on this Fund will be considered in later chapters, however it is salient to recognise that the Prospectus of the Community Housing Fund virtually repeats the definition of 'community-led housing' as given in the definitions above. The Prospectus stated that proposals for funding would be considered if they ensure that:

- meaningful community engagement and consent occurs throughout the development process (The community does not necessarily have to initiate and manage the process, or build the homes themselves, though some may do);
- the local community group or organisation owns, manages or stewards the homes and in a manner of their choosing;
- the benefits to the local area and/or specified community must be clearly defined and legally protected in perpetuity.[11]

## Core motivations

To understand the kinds of core motivations that drive contemporary activities forward, information can be culled from a combination of information sources. These will include experiences from members of the original Mutual Housing Alliance and the current Community-Led Homes partnership; data from other comparable or associated UK bodies; national and international academic reports and different 'advice notes'[12] on community-led, collaborative or 'self-organised' housing projects; and views that come forward from other national and international networks linking local communities and their community developments.[13]

Table 1.1 brings this information together to summarise what can be identified as driving people to create or shape their own homes and neighbourhoods in the UK. This summary

**Table 1.1: Key drivers behind local people creating their own homes in the UK**

| |
|---|
| • to build new property for individuals, groups or neighbourhood organisations |
| • to extend choice and variety within local housing provision |
| • to provide affordable housing, for rent or sale, for allocation to local people |
| • to have a collective or communal ownership of dwellings by the residents |
| • to be influential over local housing management and maintenance services |
| • to meet a change in household needs (like 'downsizing' for senior residents) |
| • to renovate and rehabilitate under-used or vacant property |
| • to create sustainably constructed 'low-impact' dwellings |
| • to live in 'intentional communities' (or groups), sharing similar values |
| • to establish homes within collaborative neighbourhoods |
| • to use flexible forms of movable or 'non-settled' accommodation |
| • to secure the stewardship of local assets by 'community' organisations |
| • to sustain social dynamics within existing settlements and neighbourhoods |
| • to stimulate wider opportunities for local skills and employment |

demonstrates some key ambitions of local people to shape local housing and neighbourhood circumstances:

- to have a deciding role in the creation of the dwellings that will be most appropriate for local households, particularly on the design, cost and tenure of those properties;
- to acquire and manage empty or 'under'-used property in order to give it a suitable and positive use;
- to influence how housing resources are allocated and properties are maintained;
- to shape the design and facilities of entire neighbourhoods, not just the quality of some residential dwellings, at times

in order to provide for 'communities of interest' sharing ideological, political or social backgrounds;
- to meet a diverse range of needs of individuals and of groups, which themselves may be organised in a variety of ways and scales and be from varied cultural backgrounds.

The list above also includes aspirations less frequently included in descriptions of households seeking control over their home environments, such as the motivations of 'travelling' households, or for creating 'eco-housing' that will minimise energy consumption for low-income households.

There is lastly an inclusion of some associated community-based aspirations, such as enhancing the skills and employability of local residents, or to establish a local *stewardship* over cherished buildings, places or facilities (similar to the historic description of being a steward over the assets of a country estate).

What is not included in the summary is any reference to individuals or community agencies being motivated to undertake building or property development primarily to generate capital profit. Some local projects may certainly hope that funds *could* be generated to cover the costs of other community-centred activities (or even for further building plans), but that hope for the future is rarely sufficient reason to devote such time and energy to complex activity in the present day. The information in Table 1.1 deliberately also excludes motivations to raise profits for personal gain, even if, here and there, that may be an intention of some private individuals. The message of the list above is fundamentally what galvanises a variety of *collaborative* works.

How such motivations have subsequently informed specific practices is explored in the next chapters.

## Notes

1. Around 200 or so societies remain in the UK, offering a range of financial services products to their members.
2. These have invariably undergone a series of successive renovations such that they now resemble leafy but orthodox rural and suburban development.

[3] See the report in the Guardian online, 19 January 2020, www.theguardian.com/society/2020/jan/19/housing-giants-put-profit-before-peoples-needs-report-reveals

[4] Affordable Housing Commission (2019), 'Defining and Measuring Housing Affordability: An Alternative Approach', AHC, Smith Institute, UK.

[5] Hammond, J.L & Hammond, B. (1917), *The Town Labourer 1760–1832*, Longmans, Green & Co, London.

[6] The UK Mutual Housing Alliance included CDS Co-operatives; Community Gateway Network; Confederation of Co-operative Housing; Co-operatives UK; National Community Land Trusts Network; Community Self-Build Agency; Locality; National Custom & Self-Build Association; National Federation of Tenant Management Organisations; Radical Routes; Self-Help Housing; UK Cohousing Network.

[7] Now called 'World Habitat' (see www.world-habitat.org/publications/).

[8] Building and Social Housing Foundation (2015), 'Scaling Up Community-led Housing: A proposal to the Nationwide Foundation', Coalville, UK, p 3.

[9] Building and Social Housing Foundation (2016), 'Community-led Housing', Coalville, UK, p 2–3.

[10] Homes England (2018), 'Community Housing Fund Prospectus' (Open Government Licence).

[11] The CHF Prospectus also notes 'this may be done through a mutually supported arrangement with a Registered Provider that owns the freehold or leasehold for the property'.

[12] See the very useful summary compiled by Lang, R., Carriou, C. & Czischke, D. (2018), 'Collaborative Housing Research (1990–2017): A Systematic Review and Thematic Analysis of the Field', *Housing, Theory and Society*, 35(1): 10–39.

[13] Including National Self-Build Centre; Canal and River Trust; Community Finance Solutions; Diggers & Dreamers; Action on Empty Homes; Shelter; Global Ecovillage Network; Eurotopia; European Network for Housing Research.

## Further reading

Barlow, J., Jackson, R. & Meikle, J. (2001), *Homes to DIY For: The UK's Self-build Housing Market in the Twenty-first Century*, Joseph Rowntree Foundation, York, UK.

Barton, H. (ed) (2000/2013), *Sustainable Communities: The Potential for Eco-Neighbourhoods*, Earthscan, London.

Benson, M. & Hamiduddin, I. (2017), *Self-Build Homes: Social Discourse, Experiences and Directions*, UCL Press, London.

Coates, C. (2001), *Utopia Britannica: British Utopian Experiments 1325–1945*, Diggers & Dreamers Publications, BCM Edge, London.

Coates, C. (2012), *Communes Britannica – A History of Communal Living in Britain 1939–2000*, Diggers & Dreamers Publications, BCM Edge, London.

Colenutt, B. (2020), *The Property Lobby – The Hidden Reality Behind the Housing Crisis*, Policy Press, Bristol, UK.

Eno, S. & Treanor, D. (1982), *The Collective Housing Handbook*, Laurieston Hall Publications, Castle Douglas, Scotland.

Gooding, J. & Johnston T. (eds) (2015), *Understanding the Potential of Small-Scale Community Led Housing*, Locality, London.

Gosden, P.H. (1973), *Self Help: Voluntary Associations in Nineteenth Century Britain*, Batsford, London.

Hadfield, A.M. (1970), *The Chartist Land Company*, David & Charles, Newton Abbot, UK.

Hall, P. (1988/1996), *Cities of Tomorrow*, Blackwell Publishers, Oxford, UK.

Hall, P. & Ward, C. (1998), *Sociable Cities: The Legacy of Ebenezer Howard*, Wiley, Chichester, UK.

Hardy, D. & Ward, C. (2004), *Arcadia for All*, Five Leaves Publications, Nottingham, UK.

Hetherington, P. (2015), *Whose Land Is Our Land?*, Policy Press, Bristol, UK.

Heywood, A. (ed) (2016), *Local Housing, Community Living: Prospects for Scaling Up and Scaling Out Community-Led Housing*, Smith Institute, London.

Ospina, J. (1987), *Housing Ourselves*, Hilary Shipman, London.

The Teachers (1980), *Alternative Communities Directory for the British Isles*, The Teachers Community, Bangor, North Wales.

Thacker, J. (1993), *Whiteway Colony: The Social History of a Tolstoyan Community*, Sutton Publishing, Stroud, UK.

Turner, J.F.C. (1976), *Housing by People*, Marion Boyars, London.

Ward, C. (1976/1983), *Housing: An Anarchist Approach*, Freedom Press, London.

Wallace, A., Ford, J. & Quilgars, D. (2013), *Build-it-Yourself: Understanding the Changing Landscape of the UK Self-build Market*, Centre for Housing Policy, University of York, UK.

Yallop, J. (2015), *Dreamstreets: A Journey through Britain's Village Utopias*, Jonathan Cape, London.

# TWO

# Models and practice

Some historic terms for practices found within the UK's community-led and self-build sector have already been noted in Chapter One. This chapter unpicks how the motivations underpinning local and community-led initiatives can be understood to have informed the kinds of activity that have been used to achieve particular ends.

This has been most evident in the kinds of *practices* or *models* that local projects have chosen to use for their engagements. Table 2.1 summarises the appeal of a range of 'models' or 'typologies' of local practices that have featured in the creation of local homes and neighbourhoods. These include both collaborative projects and individualistic activities.

It should be noted there is no mention here of more mainstream housing bodies or practices – bodies such as local charities, or Housing Associations, or local authority housing departments (that is, 'council' housing), or even almshouses – all of which at times can describe or present their activities as being 'community-based'. Certainly, these may involve local communities and can provide invaluable housing and neighbourhood services in their own ways, but the usual decision-making structures of such bodies are invariably dominated by management or executive bodies within very

**Table 2.1: Models of community-led and self-organised housing practice**

| |
|---|
| **Self-build housing:** Tailor-made or designed properties, arranged by individuals or groups for their own use, including homes built by the residents themselves |
| **Custom-build housing:** Where households have made use of a specialist provider/developer to finalise the 'customising' of homes to their own choosing |
| **Co-operative/mutual housing:** Housing that is owned and/or controlled by the democratic membership and engagement of all the resident households |
| **Tenant management organisations:** Tenants and leaseholders being collectively responsible for managing local homes and housing services |
| **Self-help housing:** Bringing empty or derelict properties back into use through renovation works undertaken by community projects |
| **Community land trusts/development trusts:** Housing and other assets being owned or managed for community benefit, and at permanently affordable costs |
| **Cohousing:** Creating mutually supportive neighbourhoods that combine self-contained dwellings with other shared spaces, buildings and facilities |
| **Low-impact housing:** The design and development of dwellings to maximise the protection of the local and natural environment |
| **Intentional communities:** Sharing lifestyles and properties alongside others holding common social, political, or spiritual values |
| **Homes to travel with:** Using mobile dwellings or other temporary settings, instead of homes in more permanent places |
| **'Community anchors':** Organisations securing the long-term stewardship of buildings, land and other facilities for the benefit of local communities |

professionalised and highly organised systems. It would not be the case that such structures and their services would fit with the community-led housing that has been described above. The typical scale and nature of the decisions being enacted by such executive decision-making bodies is invariably different from the local scale of collaborative or self-managed projects being depicted here; for that reason they are not included in Table 2.1.

It would also be fair to point out that several community-led housing projects have arisen *in opposition to* the policies and practices instigated by some mainstream bodies, sometimes as a response to a perceived lack of practical accountability by elected representatives or benefactors to their host communities. This is particularly the case for some tenant management initiatives, when local tenants have campaigned for a greater influence over the local services supplied by a housing association or a local authority housing department.

The main sections below are a substantial exploration of the different models listed in Table 2.1 and a basis for understanding how these can link with the intentions and definitions listed in Table 1.1. Detailed information is set out on a comprehensive range of practices and activities through which community-based housing initiatives take place and underlying motivations could be met, including:

- definitions of key terms (in the main provided by community-led organisations);
- notes on the key appeal and motivations behind each practice;
- comment on the recent context and examples of each practice;
- notes on relevant organisations offering support and advice;
- some points on implications for future policy.

The thrust of each section is to provide a full description of what different practices do, and their main context(s). Broad descriptions and the location of individual examples are provided, plus details of associated bodies of interest. There has been no attempt to compile detailed case studies as many already exist on websites and in the publications and reports from the community-led and academic/research sectors.

Where projects are named, contact details are given so that further information can be sought, if desired.

Finally, some short summaries are included to suggest how the success of projects can be justly interpreted as examples of how the application of different practices have fruitfully connected with underlying drivers and motivations.

## 2.1 SELF-BUILD AND CUSTOM-BUILD HOUSING

The classic notion of this is households arranging the creation of their own homes. The terms self-build and custom-build denote two complementary things: (a) tailor-made or designed properties, arranged by individuals or groups for their own use, including homes built by the residents; (b) households making use of a specialist provider or developer to finalise the 'customising' of homes to their own choosing.

---

### Clarification of terms

- Self-build is the term that has been used to denote 'housing projects where someone directly organises the design and construction of their new home' (National Custom and Self Build Association).

  The most obvious example is a traditional 'DIY self-build' home, where the self-builder organises the design wanted and then personally undertakes much of the actual construction work involved' (National Custom and Self Build Association)
- Custom-build has become a term being used to denote 'homes built through households working with a specialist developer to design and deliver the final unit(s)'. (National Custom and Self Build Association)

  A legal definition of self-build and custom housebuilding in the Self-build and Custom Housebuilding Act 2015 (as amended by the Housing and Planning Act 2016, www.legislation.gov.uk/ukpga/2015/17/contents/enacted/data.htm) states:

  > 'self-build and custom housebuilding' means the building or completion by:

(a) individuals,
(b) associations of individuals, or
(c) persons working with or for individuals or associations of individuals, of houses to be occupied as homes by those individuals.

The guidance notes that a minimal 'tweaking' of dwellings principally designed and built by property speculators will not meet the 'custom-build' description.

- Community self-build is the term that has been used to describe 'a group of local people in housing need, building their own homes ... Those without building skills, participate on the basis that they are prepared to make the necessary time commitment and be prepared to learn new skills'. (Community Self Build Agency).
- Group self-build usually refers to 'a group of individuals coming together to build the homes together by pooling their resources' (Community Self Build Agency).
- The Self Build Register/Right to Build Register stems from a legal duty on every local authority in England to maintain a register of people who are looking for a serviced plot to build their own home and to grant ongoing development permissions to meet the demand identified within three years.

## Key appeal (self-build and custom-build)

The attraction of being able to 'build' a property for one's own household is an ideal clearly shared by many. National surveys have consistently shown over 50% of respondents aspire to undertake such a venture at some time in their lives (e.g. from the Building Societies Association, 2011). The flexibility of organising or undertaking the construction of one's own property has enabled many households to build in very singular locations, and to very personal specifications. It has also been used by community projects seeking to maximise local participation and neighbourhood impact.

The National Custom & Self Build Association (NaCSBA) has listed the following variations (and the respective levels of 'hands-on' effort required) in how such schemes might take place:

*Self-build home*: a household manages the design and construction process of a 'one-off' property and undertakes a fair proportion of the actual building work.

*Contractor-built home*: a household arranges the design of a 'one-off' property and then selects a contractor to complete all the construction work.

*Kit home*: a household selects a type of home (and may organise the foundations) and engages the kit home company to provide and erect the property to completion.

*Independent community collaboration*: a group of households acquires a site, divides it into plots, then each household undertakes the design and construction of their home.

*Supported community self-build group*: a body like the CSBA (below) or a local community-led trust supports a group of households to build homes together.

*Developer-led home*: A developer with a site and a prearranged design erects the building for a household to customise or finish aspects of the dwelling and interiors.

*Developer-led group custom-build*: A developer builds homes for a group, either customised to completion, or for the households to complete, singly or together.

In some cases, the generic term 'self-build' is used as shorthand for a variety of the activities listed above and may also be applied to work undertaken to renovate existing properties, either by an individual or by a group. Firms and contractors supplying 'custom-build' services have felt, however, that their identity and role may not have sufficient presence in such a shorthand use, or reflect how the 'DNA' of the term custom-build reflects the routine experience of households looking to customise established designs or properties to their own needs. In 2014, therefore, the national promotional and support body changed its name from the 'National Self Build Association' to the 'National Custom & Self Build Association' – NaCSBA.

## Recent context (self-build and custom-build)

The annual UK output of the self-build/custom-build sector has been calculated to be in the region of 8–10%. This is not an insignificant number – effectively it's about the same number of homes as are built by any one of the UK's largest volume housebuilders. But precise figures are notoriously difficult to compile, and there is little data that separates figures out for the individual countries in the UK.

The NaCSBA has a target to facilitate the doubling of the overall national rate to 16% in the next few years, while market reports estimate a 41% growth in value of the sector by 2020.[1]

Government endorsement of such ambition gave support for a National 'Self Build Action Plan' produced in 2011 by the (then) National Self Build Association and other stakeholders to promote ways for self-build housing to become a 'more mainstream housing option' and increase its yearly outputs. Continual lobbying of the issues both inside and outside of Parliament, including by a very supportive All Party Parliamentary Group (now termed the All Party Parliamentary Group on Self-Build, Custom and Community Housebuilding and Placemaking) subsequently laid the ground for the Self-Build and Custom Housebuilding Act 2015 and regulations on local authorities to establish Self Build Registers and identify land available to meet local self-build and custom-build aspirations. All authorities now have a 'register' (of some kind or the other) and record 'demand' for households wishing to build individually or in groups, but little formal practice is yet to emerge on how authorities are bringing actual land into the use of those households whose demand is now on record.

Informal evidence of self-build opportunities outside of England suggests that self-building still has routine appeal as a chosen activity in the rural regions of Ireland, Wales and Scotland (the writer recalls an estate agent's window in Aberdeenshire during 2013 that was completely covered in self-build plots offered for sale), although the Scottish Government has recognised new entrants to such a market may need assistance in some locations. A two-year pilot self-build loan fund took effect from 2016 in the Highland Council area, and some useful case studies have been compiled by the Highlands Small Communities Housing Trust.[2]

Since 2017, a Right to Build Expert Task Force has been operating with support from the Ministry of Housing, Communities and Local Government 'to help local authorities, community bodies and other organisations across the UK deliver large affordable custom and self-build housing projects'. It has received funds from the Nationwide Foundation and has established a Board that includes input from World Habitat, the National Housing Federation, Royal Institution of Chartered Surveyors and the Royal Town Planning Institute.

## Examples (self-build and custom-build)

Examples of completed projects demonstrate the breadth of homes that can feature under the self-build/custom-build 'banner'.

*Self-build home*: see the annual examples of award winners in a variety of 'categories' listed from *Build It* magazine (https://builditawards.co.uk/); the 'Field of Dreams' at Findhorn (www.ecovillagefindhorn.com).

*Contractor-built home*: the myriad of different projects shown on Channel 4's *Grand Designs* programme(www.channel4.com/programmes/grand-designs); units that are being erected on the Graven Hill development at Bicester (https://gravenhill.co.uk).

*Kit home*: there is an abundance of kit suppliers operating in the UK and across Europe, including Carpenter Oak (http://carpenteroak.com); Potton Self Build Homes (www.potton.co.uk); Huf-Haus UK (www.huf-haus.com/uk/the-huf-house).

*Independent community collaboration*: the most accomplished recent example is at Ashley Vale, Bristol (www.selfbuildportal.org.uk/ashley-vale).

*Supported community self-build group*: the CSBA's recent projects, including those with ex-service personnel (www.communityselfbuildagency.org.uk/); St Minver Community Land Trust in Cornwall (www.the-self-build-guide.co.uk/st-minver-community-land-trust.html).

*Developer-built 'one-off home'*: see the database that is continually being updated by NaCSBA (https://selfbuildportal.org.uk/); Graven Hill, Bicester.

*Developer-led group project*: K1 Cambridge Cohousing (www.cambridge-k1.co.uk); Urban Splash (www.urbansplash.co.uk/residential/new-islington).

### Policy implications for self-build/custom-build projects

- Most self-build and custom-build projects in the UK are currently undertaken by households that already have access to some financial resources (which may principally be in the equity of the home they already inhabit). It would be a huge advance for households currently faced with rising housing costs fast outstripping their incomes (many currently forced to use the private rental sector) if they could be assisted to redirect the use of their incomes into affordable self-build initiatives.
- The number of 'group' schemes taking place is still less than projects for one-off buildings. There remains a crucial need to develop the skills of those wishing to build or organise collaboratively, along with the skills of any 'enabling' supporters to help schemes steer their own course. Self-build or custom-build construction may, of course, be a design and construction route adopted by other kinds of group projects, like a cohousing or community land trust scheme.
- The portrayal of the 'custom-build' concept as essentially a relationship with a housing or property 'developer' (evident in the terminology of the 2017 Housing White Paper) may reinforce an assumption that a 'developer' must be engaged in such activity. (For example, a household could engage a contractor (i.e. not a 'developer') to undertake some first stages of construction that the household will complete.) It is possible that the mainstream 'development sector' could use the

political interest in 'custom-build' possibilities as another opportunity to massage the public perception of its business models. If the end purchasers in custom-build schemes are not able to benefit from some of the savings regularly achieved by other self-build projects, then this may prove a lost opportunity to build new property at reduced costs.

---

## Key organisations and information (self-build and custom-build)

- The National Custom & Self Build Association (NaCSBA; www.nacsba.org.uk) is the 'voice of the custom and self-build sector' set up to raise the awareness of the self-build and custom-build sector; to collate and communicate best practice of the sector; and to lobby government, landowners and finance providers to help increase annual contributions by the self-build and custom-build sector to the number of homes built.
- The NaCSBA has established the Right to Build Toolkit (http://righttobuildtoolkit.org.uk) as a website providing practical guidance and information for councils, housing associations, community organisations and builders/developers that want to facilitate custom and self-build projects across the UK. It explains some of the build models supported and enabled by local councils and the private sector – those aimed towards individuals and those geared towards group projects – and provides other guidance in the form of Briefing Notes.
- The Self Build Portal (www.selfbuildportal.org.uk) is an information portal established by NaCSBA, and endorsed by government departments, to provide advice and details on all matters pertaining to custom-build and self-build development. It carries information on plots, suppliers, events, case studies and other news on national and local policy developments.
- The Right to Build Task Force (https://righttobuildtoolkit.org.uk/right-to-build-task-force/#) - is an independent expert task force launched, with endorsement from the

government to facilitate delivery of affordable custom-build and self-build schemes. It plans to work with at least 80 local authorities, community bodies and partners, and the idea is that others can benefit through regional events and case studies on the Right to Build Portal.

- The National Self Build & Renovation Centre in Swindon (www.nsbrc.co.uk) is the UK's main permanent venue designed to meet the needs of anyone looking to undertake a self-build, renovation or home improvement project. It has full-size exhibition houses to visit, as well as educational areas featuring an array of displays covering everything from groundworks and foundations, to building systems and roof types, and provides advice for visitors, as well as offering services and products to aid each step of a building project.
- The Community Self Build Agency (CSBA; www.communityselfbuildagency.org.uk) is a charity whose aim is to help people in housing need to build their own homes. It facilitates projects for men and women from all backgrounds to benefit individually and collectively from being a member of a self-build housing group, and thereby to acquire a range of skills and experience. The Agency is particularly keen for those with limited incomes or who are unemployed to participate in building schemes. Key support has been provided in recent years to a number of schemes where the training and construction work has been targeted towards assisting people leaving the armed forces.
- The Walter Segal Trust (www.segalselfbuild.co.uk/home.html) was originally set up to promote the Segal technique of simplified timber-frame housebuilding, and its applicability for households with limited or zero previous building experience, working with a variety of group and individual self-build projects to promote timber-frame techniques and build ecologically sound dwellings. Nowadays the trust is almost solely a repository of details about those first Segal buildings that still attract visitors from far afield. There is an email address for enquiries.

- The Planning Portal (www.planningportal.co.uk/info/200130/common_projects/49/self_build_homes) is a national website for planning and building regulations information, with step-by-step information on the self-build process.
- The Self Build Guide (www.the-self-build-guide.co.uk) is a website with information and hands-on experience from other self-builders, plus pointers towards products, services and suppliers.

The growth of services geared towards households creating their own self-build or custom-build home remains evident in the prevalence of the 'Self Build' and 'Grand Design' roadshows and exhibitions around the country. Information can regularly be obtained from the following:

- National Self Build and Renovation Show (http://national.homebuildingshow.co.uk).
- Futurebuild (www.futurebuild.co.uk/about; NB – this is the successor of Ecobuild).
- Grand Designs Live (www.granddesignslive.com).
- in the pages of key magazines: *Self Build & Design* (https://pocketmags.com/selfbuild-and-design-magazine); *Grand Designs* (www.granddesigns.magazine.co.uk); *Build It* (www.self-build.co.uk); *Homebuilding and Renovating* (www.homebuilding.co.uk).

## Further reading

AMA Research (2016), 'Self-Build Housing Market Report: UK 2016–2020 Analysis', www.amaresearch.co.uk.

Benson, M. (2014), 'Creating a Nation of Selfbuilders: Interim Project Report', Department of Sociology, Goldsmiths University, London.

Borer, P. & Harris, C. (1998), *Out of the Woods: Ecological Designs for Timber Frame Self Build*, Centre for Alternative Technology Publications, Machynlleth, Wales.

Brindley, M. (2017), *The Housebuilders Bible, Edition 12*, Ovolo Publications, Milton Keynes, UK.
Broome, J. (2007/8), *The Green Self Build Book*, Green Books, Totnes, Devon, UK.
Community Self Build Agency (1994), *Building Lives by Building Homes*, CSBA, Bristol, UK.
Highlands Small Communities Housing Trust (2016), 'Highland Self Build Loan Fund (HSBLF) Fact Sheet', selfbuild@hscht.co.uk.
Matthews, R. (2002), *All About Selfbuild: A Comprehensive Guide to Building Your Own Unique Home*, Blackberry Books, Godshill, Isle of Wight, UK.
National Custom & Self Build Association (2011), 'An Action Plan to Promote the Growth of Self-build Housing', Swindon, UK.
Ospina, J. (1994), *Self-Build for Rent: Choice, High Standards and Affordability*, Co-operative Housing in South-East London, London.
Scottish Government (2016), *Self and Custom Build Housing Pilot Challenge Fund Prospectus*, Scottish Government, Edinburgh, Scotland.
Three Dragons (2016), 'Research Paper: Self Build and Custom Build Housing', http://three-dragons.co.uk.
Wilson, W. (ed) (2017), 'Self-Build and Custom-Build Housing (England)', UK Parliament Briefing Paper No.06784, House of Commons Library, London.
Wilson, W. & Heath, S. (eds) (2014), 'Self-Build and the Custom Build Housing Sector', UK Parliament, Briefing Paper No. 06784, House of Commons Library, London.

## 2.2 CO-OPERATIVES, 'MUTUALS' AND TENANT MANAGEMENT BODIES

The 'co-operative' ethos has been hugely influential in the history of political change in the UK and has been a core determinant for shaping egalitarian ways that people and organisations might operate together. It has been crucial in creating property that is owned and/or controlled by a democratic membership and engagement of all the resident households, including where tenants and leaseholders are collectively responsible for managing local homes.

## Clarification of terms

- A co-operative is 'an autonomous association of persons united voluntarily to meet their common economic, social, and cultural needs and aspirations through a jointly-owned and democratically controlled enterprise' (International Cooperative Alliance).
- A housing co-operative is 'a body that owns and manages its homes in a completely democratic manner', usually identified by two common factors:
  - a membership where those who live in the homes are required or are encouraged to become members – in some cases alongside other local community members;
  - that the community membership democratically controls the housing organisation in some way through general meetings of the organisation (Confederation of Co-operative Housing).
- Primary co-ops are individual and autonomous co-operative organisations controlled by their internal membership bodies.
- Secondary co-ops are bodies established by active collaboration between primary co-operatives (at times in tandem with other supporters, such as local housing associations) to promote, service and sustain the operations of housing co-operatives through advisory, training, management and development services.
- A Tenant Management Organisation is 'a means by which council or housing association tenants and leaseholders can collectively take on responsibility for managing the homes they live in' and manage other associated neighbourhood services. (National Federation of Tenant Management Organisations).
- Arms-length Management Organisations (ALMOs) are housing management bodies established to give tenants more influence in how local authority housing estates are managed (National Federation of Arms-Length Management Organisations).
- 'Mutual' bodies – 'fully mutual' co-operatives are bodies where only tenants or prospective tenants can be members and control its affairs (Radical Routes).
- The Community Gateway model is for 'tenant-controlled membership-owned organisations, operating democratically on a large scale', designed for council housing stock transfer that allows tenants to take greater control over their homes and neighbourhoods communities (Community Gateway Association).
- Model rules are governing documents set out in the form of 'template documents' that can be used by new organisations to be established on the appropriate legal footing (Co-operatives UK).

## Key appeal (co-ops and TMOs)

The percentage of co-operative housing that exists in the UK is about 0.6% of the housing stock – that is, just under 10,000 co-operatively owned and managed homes (this could be contrasted with around 1,600,000 managed by the housing association sector). The attraction of 'co-operative' housing is the transparent way the co-operative body holds the title to all property and dwellings used by its members. The maintenance and repair of properties is undertaken with a focus on maintaining the homes to excellent standards, not to make a financial profit. Property is a common asset for all members, provided through rental agreements and tenures, and not open for purchase by any individuals.

Co-operatives are incorporated as bodies that usually require their tenant members to purchase a nominal share – invariably for £1. These are not, however, tradable for a monetary figure (future sale will only be a redemption back to the co-op body), and even where a co-operative's rules may permit members to have more than one share, a co-op's core principle for governance will be that of 'one member, one vote'.

The long-established *Co-operative Principles,* as defined and revised by resolution of the Centennial Congress of the International Co-operative Alliance in September 1995, are based on the values of self-help, self-responsibility, democracy, equality, equity and solidarity. These are consciously rooted in values of honesty, openness, social responsibility and caring for others, and are traditionally set out within the following guidelines on how co-ops should operate:

1. *Voluntary and open membership* (open to all persons willing to accept the responsibilities of membership).
2. *Democratic member control* (members actively participate in setting policies and making decisions).
3. *Member economic participation* (contributing equitably to the democratic control of the co-op's capital).

4 *Autonomy and independence* (as organisations set up to be controlled by their members).
5 *Education, training and information* (for co-op members, elected representatives, managers and employees).
6 *Co-operation among co-operatives* (working together through local, national, regional and international structures).
7 *Concern for community* (working for the sustainable development of their local communities).

## Recent context (co-ops and TMOs)

The evident appeal of co-operative and other tenant management arrangements lies in the direct control over the stock that is available to their members, and in controlling the terms on which accommodation is provided. Tenancies established years ago under co-op agreement and development remain amongst the lowest-cost tenancies in the UK's housing sector.

Where the historic formal incorporation of housing co-operatives used to be as an industrial and provident society, registered with the Registrar of Friendly Societies, modern registration is operated through the Financial Conduct Authority (FCA) under the requirements of the Co-operative and Community Benefit Societies Act 2014. New bodies can register either as a 'co-operative' (principally for the benefit of its members) or as a 'community benefit society' (run primarily for the benefit of the community at large but functioning along classic co-operative lines).[3]

Some co-op bodies may decide that they do not wish to be, or to stay, registered with the FCA – for example, bodies established under some of the investment 'loanstock' arrangements promoted by Radical Routes (see below). A move to de-register can have other tax implications, so individual co-ops do need to be sure they are procuring the right advice on the pros and cons of their situation.

Co-operatives and other tenant management organisations can operate at massively varied scales: some are a single building, based upon 'shared living', with individual bedrooms or bedsits using shared kitchens and bathrooms; some are based on providing self-contained family accommodation, operating as registered independent local housing associations in their own right. Some tenant management organisations now own or manage *thousands* of properties, that encompass all of a local authority area's previous council housing stock and include many leaseholders as members alongside the rental tenants.

Over time, the core co-operative philosophy has been applied to the work of bodies set up to undertake a wide range of purposes: bodies to manage stock held by others; 'short-life' bodies established to renovate disused buildings and to regularise other arrangements that had started as 'squats'; proposals for new forms of 'mutually owned', affordable, low-cost homeownership; 'self-build' projects that erect dwellings collaboratively for each member's benefit; and companies 'limited by guarantee' for wider social benefit, based on the principle of 'one member, one vote'.

Direct accountability of a body's activities back to its membership is crucial here, such as input from members and supporters to generate finance towards potential new developments through members' 'loanstock' arrangements. Some housing associations may class themselves as 'mutual associations', but these are now basically historic terms, given that they are no longer bodies controlled by their individual resident households. The report of the Mutual Governance: Mutual Outcomes project to review 'mutuality and accountability in the housing association sector' acknowledged that 'mutuality, and the accountability it develops, needs to be part of the fabric and culture of the association, interwoven with all aspects of governance and service delivery and constantly reviewed and improved'.[4]

The rising cost of UK accommodation and perpetual concerns about the quality of accommodation available have also prompted a drive to establish Student Co-op Homes in the UK[5] (inspired by North American student co-operatives). The first examples are in Edinburgh, Birmingham and Sheffield.

A 'Housing Commission' led by Croydon Council has been established by the Co-operative Councils' Innovation Network (local authority bodies giving dedicated support to co-op principles), with a practical focus to encourage co-operative, community-led solutions to current housing problems. Its report on how councils are enabling and supporting community-led housing covers case studies from local authorities in England, Scotland and Wales, and is supporting the development and implementation of a 'technical toolkit' for local authorities.

Under the Right to Manage Regulations 2012, introduced by the Coalition Government's 'Localism' programme, a body organised by local council tenants has a right to take over the management of local housing services if a suitable management agreement can be arranged between them and their local housing authority.

The wish to have greater control over one's domestic accommodation is also a factor in in the kind of activities undertaken by households living in places traditionally seen as 'temporary', like static caravan parks and 'park homes'. Site management standards and service problems are routinely challenged by members of the National Association of Park Home Residents and the Independent Park Home Advisory Service,[6] two bodies advising 'park home' tenants on how to respond to problems with landlords and sites.

## Examples (co-ops and TMOs)

- CDS Co-operatives (CDS; www.cds.coop) – the trading name of the Co-operative Development Society Ltd – is the largest secondary housing co-operative in South East England and provides administrative, management and development support services to tenants, leaseholders and client co-operative members in 50 independent housing co-operatives. It manages almost 3,500 homes and is a registered social housing 'provider' in its own right.

- Northwest Housing Services (www.nwhousing.org.uk/) was formed as a mutual, not-for-profit company in 2006. It is owned and controlled by its customer base of a consortium of 38 independent organisations to provide professional services to housing co-operatives, small housing associations, leaseholders and social enterprises.
- The Catalyst Collective (www.thecatalystcollective.com) is a worker's co-operative, limited by guarantee, 'set up specifically to help people create a co-operative life'. Catalyst has supported a variety of housing co-ops and worker co-ops in recent years, plus other social enterprises, ethical businesses and charities.
- Co-op Homes (www.coophomes.co.uk) owns around 300 properties in London and the South East and acts as a managing agent for other community housing organisations and housing co-ops. It is a registered housing provider with the Homes & Communities Agency (now Homes England), with its own charitable status.
- Coin Street Community Builders (CSCB; http://coinstreet.org) develops and operates social housing and commercial development on London's Southbank. It has over 200 dwellings, with a mix of 1-, 2- and 3-bedroom flats, and 3-, 4- and 5-bedroom houses. It is run collaboratively by four fully mutual housing co-operatives and their combined commercial body.
- Redditch Co-operative Homes (https://rch.coop) manages nearly 300 properties in Redditch to suit varied needs, from young single people and families, to older retired people. Originally part of the Accord Group, RCH is now a division of Accord Housing Association, a single body offering improved consistency in services, communication and value for money.

- The Community Gateway Association (www.communitygateway.co.uk) in Preston and Rochdale Boroughwide Housing (www.rbh.org.uk) are two large-scale mutual housing societies. Each was set up to receive the transfer of a local authority's entire housing stock into local tenant control and ownership – 13,500 properties in Rochdale and about 6,100 in Preston. The Community Gateway Association was judged the UK's top housing association in the 2017 list of 'top 50 social landlords', compiled by *24 Housing* Magazine.
- The Co-operative Living Freehold Society (www.uk.coop/directory/co-operative-living-freehold-society-limited) owns the freehold to properties which are leased to housing co-operatives, and is registered as an industrial and provident society with not-for-profit and common ownership rules.

---

### Policy implications for housing co-ops and TMOs

- The promotion of individual homeownership models through extending 'right to buy' opportunities to tenants across the social housing sector represents a clear threat to the ownership of co-operative housing remaining as a common or 'mutual' ownership between its members.
- The direction of government policy to utilise housing assets within the social housing sector to help construct new housing will challenge existing co-ops and their tenants to use the worth of established assets towards new development.
- There is a move by local authorities and housing associations to reverse previous tenant management agreements and bring stock property back into the control of the original stockowner. Some enforced removals of tenants have occurred simply to sell property judged to be of high net worth in the open market.
- Some co-operatives and tenant management bodies need to refresh their management and governance abilities to demonstrate to regulatory bodies that they are competent to manage stock effectively and that they have instigated the changes required by revised regulations and asset management policies.
- At the current time, there is no national means of providing a unified 'voice' for tenants within the housing sector. Recent discussions between representative bodies and the government have yet to agree the manner in which tenants' concerns and issues can be dealt with at a national level.

---

## Key organisations (co-ops and TMOs)

- The Confederation of Co-operative Housing (www.cch.coop/) has been operating since 1992 as the main national representative body for housing co-operatives and other tenant-controlled housing organisations and federations. It presents a consensus of views on how co-op and tenant management housing can operate; provides substantial services to its members; and undertakes consistent lobbying as a founding member of the recent Community Led Homes Alliance.
- Radical Routes (www.radicalroutes.org.uk/) is a member-based organisation of housing co-ops, several worker co-ops, a couple of social centres and other supportive individuals dedicated to working 'for positive change' to take control of housing, education and work opportunities.
- The London Federation of Housing Coops (LFHC; www.lfhc.org.uk) is a body with a 30-year history in London providing training and support for housing co-ops in the capital and a point of contact between other co-ops and organisations of tenant-controlled housing.
- Co-operatives UK (www.uk.coop) is the network for Britain's thousands of co-operative businesses. Co-operatives UK works to promote, develop and unite the member-owned businesses worth £36 billion to the British economy in 2017.
- The National Federation of Tenant Management Organisations (NFTMO; www.nftmo.com) was founded in 1992 to represent tenant management co-ops, estate management boards and other forms of tenant management organisations in England, with a range of support for member TMOs and other tenants.
- The National Federation of Arms-Length Management Organisations (www.almos.org.uk) is the UK body representing arms-length management organisations to develop services for local residents in partnership with local councils.

- TPAS (www.tpas.org.uk) – once termed the Tenants Participation Advisory Service – has been promoting, supporting and championing tenant involvement and empowerment across the England for over 25 years, and provides training and accreditation schemes that support resident involvement arrangements.
- The TAROE Trust (www.taroetrust.org.uk) is a charity formed in 2013 from the former national tenant representative body, Tenants and Resident Organisations of England (TAROE), to continue work on housing opportunities for socially and economically disadvantaged communities.
- Co-operatives Housing International (www.housinginternational.coop) is a sectoral organisation of the International Co-operative Alliance (https://ica.coop/) established to promote the development of co-operative housing in all countries.
- Housing Europe (www.housingeurope.eu) is the European Federation of Public, Co-operative and Social Housing. It operates a network of 45 national and regional federations bringing together 43,000 housing providers from 24 countries, managing over 26 million homes, 11% of existing dwellings in Europe.
- Social Enterprise UK (www.socialenterprise.org.uk) is the principal body in the UK for 'businesses with a social or environmental mission'. Its State of Social Enterprise Survey 2017 noted that 6% of UK social enterprises had housing as their 'principal' trading activity.

## Further reading

The 'key' bodies cited above all have significant lists of publications and other forms of advice. See also:

Acorn Co-op Support (2016), *Feasibility Study for a National Body of Student Housing Co-ops*, Students for Co-operation/East of England Co-op Society, https://www.students.coop/.

Birchall, J. (2013), *Building Communities: The Co-operative Way*, reissue of 1988 edition, Routledge Revivals, London.

Bliss, N. (ed) (2017), *Mutuality and Accountability in the Housing Association Sector*, Confederation of Co-operative Housing, Liverpool, UK.

Bliss, N. & Lambert, B. (eds) (2017), *New Co-operative and Community-led Homes (Version 2),* Confederation of Co-operative Housing, Liverpool, UK.

Bowman, A. (2003), 'Interim Spaces: Reshaping London: The Role of Short Life Property, 1970 to 2000', PhD dissertation, University of Bristol, UK.

Catalyst Collective (2010) *How to Set Up a Housing Co-operative*, Catalyst Collective Ltd., London.

CCIN Housing Commission (2017), 'Co-operative Case Studies', www.councils.coop/publications/ccin-commission-report-community-led-housing-key-role-local-authorities/.

Commission for Co-operative and Mutual Housing (2009), *Bringing Democracy Home*, Commission/CCH, Liverpool, UK.

Confederation of Co-operative Housing (2010), 'Code of Governance for Housing Co-operatives', CCH, Liverpool, UK.

Confederation of Co-operative Housing (2012), 'The CCH Accreditation Framework for Housing Co-operatives', CCH, Liverpool, UK.

Confederation of Co-operative Housing (2016), '1,001 Co-operative and Community-led Homes', CCH, Liverpool, UK.

Confederation of Co-operative Housing (2017), 'Governance and Management: Guide for Co-operative and Community-led Housing', CCH, Liverpool, UK.

Co-operatives UK (2014), 'Fully Mutual Rules', Manchester, UK.

Gulliver, K., Handy, C. & Morris, J. (2013), *More than Markets: Mutual & Co-Operative Housing in the UK*, Human City Institute, London.

Hands, J. (1975), *Housing Co-operatives*, Society for Co-operative Dwellings Ltd, London.

The Hive (2016), 'Community Benefit Society – A Guide', www.thehive.coop.

Lambert, B. (2010), *Financing Co-operative and Mutual Housing*, CCH, Liverpool, UK.

Lang, R. & Roessl, D. (2013), 'The Governance of Co-operative Housing: Current Challenges and Future Perspectives', *International Journal of Co-operative Management*, 6(2): 8–12.

Lang, R. (2015), 'Bringing Real Localism into Practice through Co-operative Housing Governance: The Prospects for Community-led Housing in England', Working Paper 1, Housing & Communities Research Group, University of Birmingham, UK.

Local Government Association (2019), 'Engaging and Empowering Tenants in Council-owned Housing', LGA/TPAS, UK.

National Housing Federation (2016), *Briefing: Transparency and Accountability in the Housing Association Sector*, NHF, London.

National Federation of ALMOs (2015), 'Key Facts about ALMOs', www.almos.org.uk.

Rowlands, R. (2009) 'Forging Mutual Futures: Co-operative and Mutual Housing in Practice: History & Potential (Phase 1 Research Report to the Commission on Co-operative and Mutual Housing)', Centre for Urban and Regional Studies, University of Birmingham, UK.

Seeds for Change (2013), 'Legal Forms for Co-ops', www.seedsforchange.org.uk

Welsh Co-operative Centre (2015), 'Co-operative Housing: For any Community in Wales', WCC, Wales.

## 2.3 COMMUNITY LAND TRUSTS AND DEVELOPMENT TRUSTS

There is a surge in enthusiasm for local bodies that can empower local communities to be in control of commercial and residential activities and to enable housing and other assets to be owned and/or managed for community benefit, especially at permanently affordable costs to members of the local community.

---

### Clarification of terms

- A community land trust (CLT) is 'a form of community-led housing, set up and run by ordinary people to develop and manage local housing, ensuring that it remains genuinely affordable, based on what people actually earn in their area. CLTs act as long-term stewards of housing not just for now but for every future occupier' (National Community Land Trust Network).

The legal definition of a community land trust was incorporated in the Housing & Regeneration Act 2008, Part 2, Chapter 1, Clause 79:

> that the body is established for the express purpose of furthering the social, economic and environmental interests of a local community by acquiring and managing land and other assets in order -
>
> (a) to provide a benefit to the local community, and
> (b) to ensure that the assets are not sold or developed except in a manner which the trust's members think benefits the local community
>
> [And that] the body is established under arrangements which are expressly designed to ensure that -
>
> (a) any profits from its activities will be used to benefit the local community (otherwise than by being paid directly to members),
> (b) individuals who live or work in the specified area have the opportunity to become members of the trust (whether or not others can also become members),
> (c) the members of the trust control it.

- A development trust or community development trust is a longstanding UK organisational framework for community organisations

  > owned and managed by the local community ... [these] aim to achieve the sustainable regeneration of a community or address a range of economic, social, environmental and cultural issues; are independent, but seek to work in partnership with other private, public and third sector organisations and to reduce dependency on grant support by generating income through enterprise and the ownership of assets (Development Trusts Association, now 'Locality').

- An asset lock refers to 'a constitutional device that prevents the distribution of residual assets to members. The purpose of an asset lock is to ensure that the public benefit or community benefit of any retained surplus or residual value cannot be appropriated for private benefit'.[7]

## Key appeal (community trusts)

Community development trusts in the UK have been local community bodies created to support community-run businesses, own commercial property and own and manage residential dwellings. They have collectively provided a host of local services, including care services, leisure services, landscape management, apprenticeships, training and employment schemes – all types of activities that can be of benefit to their local and host communities. Some development trusts have been set up with an extensive remit to work across a significant geographical area – such as Westray Development Trust working across the island of Westray in the Orkneys.[8]

The predominant focus of the UK's recent community land trusts has been to maximise the provision of local 'affordable' housing and strengthen the connection and use of such provision within their sponsoring communities. A vital element of this desire for 'affordability' has been to set up 'asset locks' on the properties provided, whereby the costs and benefit of using this will remain available to the residents of future generations 'in perpetuity'.

The National Community Land Trust Network has noted how the first contemporary CLTs in the UK were set up on quite modest scales – many for just a few properties or just a single building – however, larger-scale developments have since emerged. CLTs have been established more in rural areas than in urban ones and have mainly focused on the creation of new rental provision, although there has been some development of dwellings for low-cost and 'affordable' ownership, and there is a desire to see the model used in urban settings, with London leading this drive.

Some recent CLTs (as with other trusts in previous years) have been established to safeguard important local amenities, like the long-term ownership of land and allotments, community businesses, pubs or other workspaces. In Scotland there has been a distinctive use of the CLT model to help local communities lobby and acquire the ownership of large areas of land that landowners were about to sell on the open market. Such actions have been strengthened through the enactment of Scotland's 'Community Empowerment Act' in 2015, which has extended the community's right to buy, making it more feasible for local communities to take over public sector land and buildings.

## Recent context (community trusts)

Development trusts have had a substantial history of engagements in the UK. The original garden cities set up community-based trusts to acquire and sustain local assets on behalf of the community and there are thriving legacy bodies that remain. The charitable and self-funding Letchworth Garden City Heritage Foundation owns and manages apartments, shops, offices and industrial units spread across the 5,500 acres of Letchworth Garden City Estate.[9]

The CLT model itself is an import from the United States, where land trusts have been widely recognised as a method of delivering permanently affordable housing since the late 1960s. In the early 2000s a combination of academics and community development professionals in England applied US ideas to innovative approaches for the provision of local affordable housing. A 'national demonstration programme' led by Community Finance Solutions, with support from Carnegie UK Trust and the Tudor Trust, set up a number of pilot projects between 2006 to 2008. In 2010 the embryonic National CLT Network was established to support the growing CLT movement and advocate on behalf of CLTs and launched as an independent charity in June 2014.

CLTs are not therefore a completely new concept in the promotion of control by local communities, but it has brought a new impetus to local communities that want to have more influence and control over local housing resources. It is moreover a format that is appealing to a diverse range of local interests, including rural parish councils, 'rural housing enablers', housing associations and new community bodies seeking to be established for this purpose.

## Examples (community trusts)

- The national body of community organisations and 'anchor' bodies and trusts, Locality, has a membership that now numbers over 600 organisations across the UK, where 'every week over 400,000 people are supported by our network of community organisations'.[10]
- At the time of compiling this publication, the National CLT Network has calculated that there are '253 Community Land Trusts in England and Wales, and including new groups forming the number is over 300 … [with] 935 CLT homes and an estimated 5000 more in the pipeline'.[11]
- The Westway Trust (www.westway.org) has been driven by communities in the North Kensington area of London for over 45 years to create local businesses, arts projects, environmental projects and a wide set of community and estate development works across this part of Kensington and Chelsea. The wider community has been focused from 2017 on responding to the tragedy of the Grenfell Tower fire.
- The Moseley Community Development Trust (http://moseleycdt.com/the_organisation), operating in the Moseley area of Birmingham, is dedicated to improving the neighbourhood 'through a combination of social, economic and environmental projects'. In 2014 it promoted and secured the adoption of a formal Supplementary Planning Document for the area from the City of Birmingham.
- Cornwall Community Land Trust Limited (www.cornwallclt.org), established in 2006 and registered as an independent Industrial and Provident Society with charitable rules in 2007, provides technical assistance, advice and development agency services for community groups and small housing associations for creating new homes and securing community assets.
- Wessex Community Assets (https://wessexca.co.uk/) is a community benefit society within the Wessex Reinvestment Trust group of not-for-profit organisations, focused on supporting communities to establish CLTs, carrying out

research and advising on the issuing of community shares for projects with social benefit.

- London CLT (www.londonclt.org) is an independent organisation, governed by its own members, that was born out of the community-organising efforts of Citizens UK in response to the need for genuinely affordable homes in the capital. It started out as East London CLT in 2007, and was instrumental in establishing the capital's first successful CLT project at the St Clements Hospital site in Mile End – turning parts of a former workhouse-cum-psychiatric hospital into permanently affordable homes, employment space and local amenities.
- Rural Urban Synthesis Society (RUSS; www.theruss.org) is an urban project in South London with over 800 members, founded upon core principles of 'affordability, sustainability, self-governance and community engagement'. Its first project is an eagerly anticipated 33-dwelling scheme in Lewisham, providing homes and other communal space, along with opportunities for self-build construction.

---

### Policy implications for land trusts and development trusts

- Some contemporary descriptions of current 'community-led housing' (CLH) read as if this format is synonymous with community land trusts per se – that CLTs are somehow the 'new norm'.[12] Care needs to be taken that not all the engagements of the 'community-led movement' are automatically considered as equivalent to the practices undertaken by land trusts/CLTs.
- The assessment of the urban CLT demonstration programme noted the wide-ranging character of the urban projects that had participated in this review, including diversity in scale, origins, diversity in the type of housing being developed (including rental, 'affordable' ownership and renovation of empty homes), and a broad diversity in potential uses and beneficiaries.
- There is a significant role still to be undertaken within communities by the original form of development trusts that have been traditionally engaged with wider issues than solely affordable housing. Canvassing local support to set up a broad 'development' trust may increase a community's opportunities to dovetail housing activity with other needs and priorities.

- Local political support for CLTs and other trusts remain vulnerable to changes in national policies. The potential extension of the Right to Buy to social housing tenants, and government's directives, such as that new developments of 'starter homes' can be exempt from legal and planning requirements to provide affordable housing, increases uncertainty over the roles of bodies previously supportive of the community housing sector. Some local councils that had previously supplied operational funding to community-managed schemes have become reluctant to commit to new local service-level agreements.

## Key organisations and information (community trusts)

- The National Community Land Trust Network (www.communitylandtrusts.org.uk) is a national charitable body supporting CLTs in England and Wales. It provides funding, resources, training and advice for CLTs and works with government, local authorities, lenders and funders 'to establish the best conditions for CLTs to grow and flourish'.
- Locality (http://locality.org.uk) is a national community development agency and network of community-led organisations working together to help and support neighbourhood initiatives. It was created by the amalgamation in 2011 of the Development Trusts Association and the British Association of Settlements and Social Action Centres.
- Community Finance Solutions (https://hub.salford.ac.uk/cfs/) is an independent research unit founded in 1999 to help empower communities to address local problems relating to land and financial inclusion. It specialises in financial and social inclusion and community asset ownership, offers independent research and advisory services to social landlords, local authorities, national government, charities and other agencies, plus innovative solutions for securing community ownership of land and models for loans to low-income households.
- Community Land Scotland (www.communitylandscotland.org.uk/) represents Scotland's new generation of 'community landowners'. It manages a total of around 500,000 acres of land, which contain homes for around 25,000 people.

## Further reading

Building and Social Housing Foundation (2005), 'Redefining the Commons: Locking in Value through Community Land Trusts', Proceedings at St George's House Windsor, 21–23 June.

Community Finance Solutions (2008), *Then We Will Do It Ourselves: A Report on Rural CLTs as a Part of the CLT National Demonstration Programme*, Carnegie Trust, Salford, UK.

Davis, J.E. & Stokes, A. (2009), 'Lands in Trust, Homes that Last', www.champlainhousingtrust.org.

Hill, S. (2015), 'Property, Justice and Reason: Reconnecting the Citizen and the State through Community Land Trusts and Land Reform', Churchill Fellowship Report, http://stephenhillfutureplanning.blogspot.co.uk.

Hoover, J. (2015), 'The Human Right to Housing and Community Empowerment: Home Occupation, Eviction Defence and Community Land Trusts', *Third World Quarterly*, 36: 1092–109.

Interreg North West Europe (2018), 'Community Land Trust Case Studies', www.nweurope.eu/projects.

Moore, T. (2014), 'Affordable Homes for Local Communities: The Effects and Prospects of Community Land Trusts in England', St Andrews Centre for Housing Research, St Andrews, Scotland.

Moore, T. (2015), 'The Motivations and Aspirations of Community Land Trust Volunteers in Somerset, Dorset and Devon', Wessex Community Assets, Glastonbury, UK.

Moore, T. (2016), 'Community-led Housing: The Evolution of Partnerships between CLTs and Housing Associations', http://blogs.lse.ac.uk/politicsandpolicy/the-evolution-of-partnerships-between-community-land-trusts-and-housing-assoc.

Moore, T. & McKee K. (2012), 'Empowering Local Communities? An International Review of Community Land Trusts', *Housing Studies*, 27 [online]: 280–90.

Moore, T, Mullins, D., He, B., Thompson, M., Claassen, C. & Archer, A. (2018), 'The Urban CLT Project Evaluation', National Community Land Trusts Network, UK.

National Community Land Trusts Network (2015), 'Our Homes Our Communities: A Manifesto for Community Land Trusts', www.communitylandtrusts.org.uk.

National Community Land Trusts Network (2016), 'CLT Organisational Health Check', www.communitylandtrusts.org.uk.

National Community Land Trusts Network (2016), 'The Community Land Trust Handbook', www.communitylandtrusts.org.uk.

Paterson, B. & Dayson, K. (2011), 'Proof of Concept: Community Land Trusts', Community Finance Solutions, University of Salford, UK.

Rural Housing Alliance (2017), *Affordable Rural Housing: A Practical Guide for Parish Councils*, RHA Publications, Tavistock, Devon, UK.

Skerratt, S. (2013), 'Enhancing the Analysis of Rural Community Resilience: Evidence from Community Land Ownership', *Journal of Rural Studies*, 31: 36–46.

Wilcox, D. (1998), 'The Guide to Development Trusts and Partnerships', http://partnerships.org.uk/pguide/fulltext.htm.

## 2.4 SELF-HELP HOUSING AND EMPTY DWELLINGS

There is a steady practice of bringing empty or derelict properties back into productive use through renovation work undertaken by local community projects. A key focus of this work across the years has been for properties to be used as lower-cost housing to meet local residential need.

---

### Clarification of terms

- Self-help housing refers to 'groups of local people bringing back into use empty properties that are in limbo, awaiting decisions about their future use or their redevelopment'. (Self Help Housing)

The properties can be 'borrowed' on a licence or a lease for a specified period or bought outright. Self-help housing groups negotiate with the owners of empty properties for their use and then organise and undertake whatever repairs are necessary to make the properties habitable.

- The definition of when a dwelling is empty is largely enshrined in Council Tax assessments. These incorporate various levels of potential tax liability depending on the length of a vacancy and will reflect diverse local authority policies.
- Both a Compulsory Purchase Order (CPO) and an Enforced Sale Order enable a local authority to take over the ownership of a property and then sell it to a third party, particularly to recover costs of work necessary to make the property safe.
- An Empty Dwelling Management Order (EDMO) enables a local authority to acquire the occupation and management of privately owned properties that have been unoccupied for 6–24 months (authority policies vary), without purchasing the property.
- 'Short Life' properties are those that are empty and have an anticipated short remaining future. (This can be due to plans to demolish the property or to some other development of the land, or are hard to let, or hard to sell, or are simply run down.)

Self-*help* housing is not a term traditionally used for self-*build* activity – activity that is invariably focused on a household's action to build a new property – although self-building can sometimes involve substantial alterations to an existing property.

- Squatting is described as 'occupying empty buildings, or land, without permission: normally by homeless people seeking somewhere to live, at least for a while' (Advisory Service for Squatters)

## Key appeal (self-help housing)

Why is self-help housing appealing? Because local communities are crucially aware of the wasted local resources that empty properties represent, and of the detrimental effect and picture of relative neglect that empty properties exert on their neighbouring areas over time. Less people are aware that it's possible to get legal possession of empty properties that can help set up local renovation projects.

The range of benefits that inspire self-help activity to make positive use of empty properties has been listed by the UK's main body in this sector, Self Help Housing:

- utilising empty or redundant buildings which owners have left unused or in disrepair, but are willing to make available for a limited period on licence or lease;
- creating a source of economical and affordable housing, or other community facilities, for people who might otherwise be homeless or in inadequate accommodation;
- raising the overall quality and appearance of a neighbourhood by removing eyesores or 'blight' created by derelict and rundown properties and reducing their attraction to those indulging in 'anti-social' behaviour;
- enabling people to take direct responsibility for securing their own housing or developing community facilities, and to challenge poor standards of stock management in the public or private rental sectors;
- providing opportunities for local people to learn practical 'on the job' building skills, plus wider organisational and management skills in how to organise and operate local organisations, how to secure funding, and how to negotiate with owners, local government officials and other agencies;
- generating new workspace for start-up businesses and local social enterprises.

Historically self-help activity has been undertaken by groups of people who cannot afford to buy or otherwise secure adequate housing, and whose housing needs are such that they are unlikely to be prioritised for an offer of a permanent tenancy by the local authority or housing association (for example, in the current climate, all sorts of young and single people, childless couples, refugees, asylum-seekers).

## Recent context (self-help housing)

Government statistics published by the Ministry of Housing, Communities and Local Government[13] (MHCLG) put the number of empty homes in England in October 2018 at 634,453. This represents a 4.7% increase on the previous year's total. Of the 634,435, 216,186 were classed as long-term empty properties (empty for longer than six months). Mapping these properties shows that more northerly local authorities have a persistently higher proportion of dwellings empty for more than six months, although there are noticeable exceptions in some southern locations.

Traditionally, local authority actions to force abandoned buildings to be brought back into use have been to make a 'compulsory purchase' of a property or bring about an 'enforced sale' of a property. (This has not always proved to be a guarantee that the sold building would be brought back into immediate use.) From 2001 lobbying of government sought to promote a new power for local authorities to take temporary possession of empty properties and force them to be rented out to tenants – a form of 'compulsory leasing' process. This new kind of possession order – the Empty Dwellings Management Order – was created by legislation in 2006, allowing local councils in certain specific circumstances to take over the management of empty residential properties with a view to agreeing with owners a plan to bring them back into occupation. Changes in the law in 2012 – the Housing (Empty Dwelling Management Orders) (Prescribed Period of Time and Additional Prescribed Requirements) (England) (Amendment) Order 2012 (SI2012/2625) – have increased the period for which the property needs to have been empty before an EDMO can be applied. It has been extended from six months to two years. The Order also provides that an authority must give the owner at least three months' notice of the intention to apply for an order.

The national campaigning body, Action on Empty Homes (previously called the Empty Homes Agency, and then Empty Homes) has worked for many years to promote action on empty and derelict stock across England and is explicitly in support of community-inspired activities. It's report *Empty Homes* in 2017 that reviewed local authority strategies to reuse empty properties notes:

> we think such strategies are more likely to be successful if they involve local residents and recognise what people think works well and less well in the area. Lasting and far-reaching improvements are more likely to be secured if local people have a genuine say in what happens. In many cases, local community-led organisations are well placed to bring empty homes back into use, alongside addressing other neighbourhood issues.

In part, this position had been informed by recent experience of central government grant finance being targeted towards community-led actions and self-help groups. Between 2012 and 2015, £50m was allocated to 110 community groups in two rounds of grants, and they were subsequently able to refurbish and reuse 1,600 homes.[14] The recent report from Action on Empty Homes, 'Community Action on Empty Homes: Using Empty Homes to Regenerate Communities',[15] provides up-to-date examples of local actions that demonstrate the worth of projects tackling the blight of long-term empty dwellings.

Funds that might be available to support self-help projects invariably have to be sought from local authorities, sometimes for use of a local authority's accumulated 'right to buy' receipts, and sometimes from available 'commuted sums' finance from Section 106 planning agreements. The cost of acquiring properties may well relate to other political priorities at the time – the transfer of Liverpool properties to the Homebaked and Granby CLTs was for £1, not least to infuse the momentum sought for an invigorated approach to area regeneration.

In contrast to a planned acquisition and renovation of underused dwellings, the squatting of derelict property can still be found, although to suggest there is an 'appeal' in undertaking such action might be to risk romanticising what drives people to it. The sight of properties being left vacant of unused for extended periods of time has clearly been one spur behind direct action focused on occupying empty property, although there may be an even stronger sense of frustration (or even desperation) when no other suitable or affordable housing option has seemed to be available.

The actual amount of squatting in the UK has been notoriously hard to assess, the government estimated in 2010 that there were 20,000 squatters at any one time.[16] The Coalition Government consulted on ways of strengthening the law in this area in 2011, and subsequently the Legal Aid, Sentencing and Punishment of Offenders Act 2012 was amended to introduce a new offence of squatting in a residential building (section 144). People squatting in a clearly residential property now face almost automatic arrest. Yet whilst the overall definitions of the law are very technical, advisory services do consider that elements are still open to debate, and that the law may not apply where:

- the squatted property is not residential;
- people are or were tenants (including sub-tenants) of the property;
- people have (or had) an agreement with someone with a right to the property;
- people in the property are not intending to live there (they may be just visiting).

This is perhaps one reason for the rise in bodies promoting the 'guardianship' of empty properties like vacant offices and warehouses by introducing temporary 'live-in' residents, permitted a limited occupancy on a form of licence.

## Examples (self-help housing)

At present there are over 50 self-help housing projects in England and Wales focused upon making productive use of empty property, varying in their organisational structures, including co-operatives, development trusts and land trusts. In size they range from those managing only a few tenancies to those managing several hundred. A few examples (see also the *Self Help Housing UK Directory*, noted below):

- Canopy, Leeds (http://canopyhousingproject.org), is a community housing project founded in inner city Leeds in 1998, and now a community benefit society. It manages homes in the Beeston, Burley and Harehills areas, many leased from Leeds City Council at a peppercorn rent, and some now bought using public grants. Properties have also been acquired from East North East Homes and Aire Valley Homes Leeds, plus housing associations such as Unity Housing and Leeds Federated Housing.
- Latch (Leeds Action to Create Homes; www.latch.org.uk) is a community benefit society with tax-exempt charitable status, established in 1989, that has refurbished derelict and rundown properties in the Chapeltown, Harehills and Burley areas of Leeds.
- Giroscope, Hull (www.giroscope.org.uk), is a housing charity and award-winning body. It has been based in West Hull for over 30 years and is part of the UK's Self-Help Housing movement. It makes the following 'Pledge' to its 250+ tenants:

  > So long as the conditions of the tenancy are met, you have the right to remain in your home for as long as you wish. Giroscope will not sell your house, or transfer ownership to another landlord so long as you are a tenant of that house. Giroscope will not borrow money against its property in a way that endangers the financial stability of the organisation and puts your home at risk.

- Phoenix Housing Co-op, London (http://phoenix housingcoop.org), is a fully mutual housing co-operative founded in 1980. Members undertake refurbishment projects, support the sustainability and stability of the coop as a whole and aid the development of new partner organisations.
- Mace Housing Co-operative, London (https://macehousing.org.uk/), was set up in 1974 as a self-help housing association by graduate students from University College to house creative individuals and homeless people. Mace now manages around 300 properties, most of them leased from housing associations and local authorities, particularly including the London Boroughs of Hackney and Camden.
- North Ormsby Community Land Trust (http://north-ormesby-clt.org.uk/news.html). Community consultations carried out by the North Ormesby Big Local Partnership in 2014 highlighted poor-quality privately rented housing and empty homes as key priorities for action. The North Ormesby Community Land Trust was established to lead on local actions, developing house renovation work, social lettings arrangements and other local management initiatives.

## Policy implications for self-help/empty homes projects

- The research reports into the use of recent government grants has confirmed that five essential ingredients are necessary for self-help housing to succeed with local housing renovations:
  - empty properties are under the control of local bodies;
  - sufficient funding is available for the community body to undertake renovations;
  - an available workforce can be established, with support for volunteers;
  - support is manifest from other local residents;
  - support is manifested by willing and experienced partners.
- In April 2017 Empty Homes published the findings of a research project that surveyed 46 local authorities with higher overall levels

or concentrations of long-term empty homes within particular neighbourhoods. The survey confirmed that concentrations of empty homes are linked to high levels of deprivation and poor standards of housing, with more residents on low incomes, and relatively low house prices. Activity by local communities willing to rehabilitate empty properties in such areas does, however, challenge assumptions that derelict properties automatically require removal or demolition.

- The terms on which empty properties can be 'acquired' through Empty Dwelling Management Orders or by compulsory purchase activities can involve punitively high valuations that hamper the property's affordability. The 2017 Housing White Paper suggested that these terms could be reviewed, but this has been stated before and led to little substantial revision. Action on Empty Homes has noted the woefully small number of EDMOs that are currently being used in the UK to take control of local empty properties, local authorities hoping instead to 'persuade' absentee owners to negotiate on the use of their properties.
- Some reports are appearing of spontaneous 'shanty' encampments around the UK,[17] in 'off-grid' settings like rundown riverside or ex-industrial areas. Such ad hoc actions by low-income and homeless households struggling to stay in the area of their choice are not usually included in any formal counts of homeless households or 'rough sleepers'.

---

## Key organisations (self-help housing)

- Self Help Housing UK (http://self-help-housing.org) is the key independent body set up to raise the profile of self-help housing and to increase the resources available. It has organised support by:
  - operating a dedicated self-help housing website and Directory of Projects, for existing self-help projects to make contact with one another and to share their experiences;
  - providing advice and information for local people interested in setting up new community self-help projects;
  - working with other community, housing and regeneration organisations to develop new projects amongst their members;

  - working with both charitable and statutory funders to persuade them to finance self-help housing projects through revenue or capital funding;
  - encouraging local authorities, housing associations and private owners to consider self-help housing options for bringing empty properties into use.
- Action on Empty Homes Agency (www.actiononemptyhomes.org/ – registered as a 'community benefit company') is the national campaigning agency in England, which aims to:
  - raise awareness of the waste of long-term empty homes;
  - research and develop ideas for bringing long-term empty homes back into use;
  - provide encouragement and advice for bringing empty homes into use;
  - campaign for changes to policy and initiatives at national and local levels.

  The Agency helps organise a yearly national 'Empty Homes Week' 'as an opportunity to organise events and put out information'.
- Habitat for Humanity, Great Britain (www.habitatforhumanity.org.uk) is an international body, partnering with corporations to secure the funds to send volunteers to build and renovate homes across the UK, building alongside professionals and tackling housing poverty.
- Squatters Action for Secure Homes (www.squashcampaign.org/) is an independent campaigning body that provides advice and statistics about notable squats and activities.
- There is also the body Property Guardians (www.propertyguardians.com) that places 'vetted' guardians in temporary occupation of empty properties (on non-exclusive licenses, paying a fee without tenancy rights). These are providing a degree of occupier protection in

property owned by housing associations, private landlords and other asset owners.

## Further reading

Action on Empty Homes (2019), 'Community Action on Empty Homes: Using Empty Homes to Regenerate Communities', www.actiononemptyhomes.org.

Building and Social Housing Foundation (2016), *Peer Exchange Journal 2015–16: Self-Help Housing in the North of England*, BSHF, Coalville.

Empty Homes (2017), *Empty Homes: Why Do Some Areas Have High Levels?*, Empty Homes, London.

London Borough of Islington (2015), 'Preventing Wasted Housing Supply: Draft Supplementary Planning Document', London.

Moore, T. & Mullins, D. (2013), 'Scaling-up or Going-Viral: Comparing Self-help Housing and Community Land Trust Facilitation', Third Sector Research Centre Working Paper 94, University of Birmingham, UK.

Mullins, D. & Sacranie, H. (2014), 'Evaluation of the Empty Homes Community Grants Programme (EHCGP) Midlands', University of Birmingham, UK.

Mullins, D. & Sacranie, H. (2015), 'Evaluation of the Empty Homes Community Grants Programme (EHCGP) Yorkshire & Humberside and the North East', University of Birmingham, UK.

Mullins, D. & Sacranie, H. (2016), 'Future Scenarios after EHCGP/ Empty Homes Community Grants Programme Legacy Project', Housing and Communities Research Group, University of Birmingham, UK.

Sandford, M. (ed) (2017), 'Evicting Squatters', UK Parliament Briefing Paper 02857, House of Commons Library, London.

Squatters Action for Secure Homes (2017), 'Repeal Section 144' Petition, SQUASH Newsletter, 1 September.

Wilson,W. (ed) (2016), 'Empty Dwelling Management Orders (EDMOs)', UK Parliament Briefing Paper 04129, House of Commons Library, London.

Wilson, W., Cromarty, H. & Barton, C. (eds) (2019), 'Empty Housing, England', UK Parliament Briefing Paper 3012, House of Commons Library, London.

## 2.5 COHOUSING NEIGHBOURHOODS

Over and above the appeal of creating one's own home is for this to be within a collaborative shaping of the wider neighbourhood and its social environment. There is a consistent interest in creating mutually supportive and communal neighbourhoods that combine self-contained dwellings alongside shared spaces and facilities.

### Clarification of terms

Cohousing is the term used to denote neighbourhoods 'created and run by their residents. Each household has a self-contained, private home as well as shared community space. Residents come together to manage their community, share activities, and regularly eat together' (UK Cohousing Network).

'Cohousing' is a 1980s translation of the Danish term 'bofællesskab' used to describe Danish settings for communal living and does not appear to have had a specific presence in the English language before that time. 'Bo' in Danish means both to 'live' and to 'house', and 'fælles' means 'in common', conveying a subtle combination of being 'housed together' and 'residing together' in a shared place – having a 'living in common'. The modern English term cohousing was created to describe 'housing that combines the autonomy of private dwellings with the advantages of communal living'. (McCamant and Durrett, 1988: p.12)

(The word cohousing is used throughout this publication for this combination of private and community facilities within supportive neighbourhoods. In recent years the use of the term 'co-housing' has emerged, which has become a vogue description for diverse kinds of collaborative or shared housing initiatives. Cohousing's distinct combination of private and common amenities, in a deliberately designed setting for a collaborative neighbourhood, is a distinct practice that is measurably different from other 'shared' endeavours. There is a confusion introduced by the 'co-' prefix that risks the original characteristic combination being under-appreciated and descriptions then misapplied.)

## Key appeal (cohousing)

Central to cohousing ideas, both in theory and practice, is that they are centred around how a collection of member-households can live in a harmonious new neighbourhood setting that they have established together. The core values and objectives are to create places in which people can thrive together in practical and neighbourly relationships that are mutually supportive and beneficial to all.

This modern response to desires for creating 'shared' or 'intentional' neighbourhoods draws heavily from McCamant and Durrett's seminal work during the 1980s and 1990s with Danish and North American communities. Whilst they noted that the original cohousing developments they visited in Denmark might have had subtle variations in size, location, types of ownership and local priorities, nevertheless they saw how the success of such neighbourhoods routinely featured essential characteristics:

- a design of the physical form and layout of the neighbourhoods to maximise a constant incidental and organised contact between residents;
- extensive common facilities and spaces supplements the private self-contained accommodation, of which a 'common house' is a crucial element: common spaces and facilities are perceived as shared extensions to one's own house and amenities;
- the size and scale of each cohousing neighbourhood is appropriate to underpinning sustainable community dynamics and relationships between all its households;
- all decisions are under the control of the residents, and it is the group that manages the co-design, co-development and co-organisation of the shared neighbourhood life.

Whilst some of these characteristics are evident in other forms of collaborative housing schemes, what has made cohousing distinct from other approaches in practice is how its intentional shared and 'participatory process' maximises the creation of a shared identity and a supportive neighbourhood life. As McCamant and Durrett note 'none of these characteristics is unique [to cohousing], but the consistent combination of all four is' (McCamant and Durrett, 2011: p.43).

On the matter of the size of cohousing projects, McCamant and Durrett noted: 'we take seriously the advice on community size from the Danes because they have a long history in making cohousing work. A cohousing community that contains 20 to 50 adults seems an optimum size' (McCamant and Durrett, 2011: p.31). This focus on the numbers of households in cohousing communities is different from the usual considerations for other UK community-led housing ideals. It stems from an awareness of the sociodynamic experiences of community settings elsewhere that have had too many households for people to know one another properly, or too few for members to be able to find a comfortable distance from any interpersonal friction that might arise. This is, however, recognised by the UK Cohousing Network, which advises that 'the size and scale of cohousing [needs to be] appropriate to support community dynamics for easy informal communal contact; this is usually between 10–40 households'. Larger settings with a cohousing character may be built, but the dwellings can then be grouped around more than one common house and shared set of amenities, as established in urban settings at Århus in Denmark. This strengthens the likelihood that the desired neighbourly dynamics can naturally develop from collections of households being geographically orientated to build up a routine use of one set of common facilities rather than another.

Yet it is also fair to note that the lack of familiarity in the UK of the kind of community dynamics that enable cohousing residents to thrive in their day-to-day lives has seen attempts to create cohousing settings that do not have all its core elements in place. Understanding the right size for such neighbourhoods is one important facet; another essential aspect is the role of a 'common house', functioning as a central point of contact and collaborative activity for all the neighbourhood's households – meaning that it should be large enough to bring all community members together at any one time (even if some activities, like sharing some meals together, are only regularly sought by a portion of the community). Some collaborative projects have expressed a view that such a facility is not essential, or may seem too expensive, or too problematic to manage, but all the evidence from established cohousing communities emphasises the essential value for engagement and contact that a common house creates.

## Recent context (cohousing)

The drivers behind almost all the British cohousing communities noted below stem from clear ambitions to mirror the vibrant neighbourhoods that have emerged in the past few decades in Scandinavia, Northern Europe and North America.

Innovative UK schemes created since 2000 have included the reuse of ex-health service sites (The Community Project), pocket sites in urban areas (Stroud, Cannock Mill and Lancaster), and extensions of land already occupied by collaborative households (East Whins and the Threshold Centre). However, examples that properly reflect the depth of interest in undertaking such neighbourhood development have struggled to materialise. Finding sites suitable for group-led developments that are bigger than most other kinds of community-led initiatives is a key obstacle here, notwithstanding that ideas for cohousing neighbourhoods can fit very well alongside the other kinds of larger sites allocated for mainstream development and urban extension areas.

Bunker and colleagues could find only a modest success rate of cohousing projects in the Diggers & Dreamers Review of 2011, compared with a continual rise in the number of groups that have ambitions to create such schemes. Some projects have only succeeded through a remarkable persistence shown by core members (and from some core supporters), and do not thereby lead to an easily replicable practice. Other obstacles hindering new projects that have been identified by the UK Cohousing Network include:

> [a] poor understanding of cohousing by local authorities; difficulty in accessing social housing grants to make cohousing affordable; high land prices; a lack of capacity in groups hoping to establish new communities; and difficulties in working with housing associations which are reluctant to allow the autonomy that is vital for cohousing communities.[18]

Real pioneering development does, however, continue to emerge. The Older Womens Cohousing group (OWCH) development in North London has demonstrated the innovative emphasis of projects steered by 'older' (that is, senior) members. This mirrors other continental examples that have consistently featured in The Housing our Aging Population – Panel for Innovation (HAPPI) reports on contemporary 'best practice' community settings for older people. OWCH, along with the Threshold Centre in Dorset, has also created a mix of tenures within the one community – a useful rebuttal of a simplistic criticism about cohousing projects that they may only suit a development of property intended for private ownership.

## Examples (cohousing)

Notwithstanding the modest pace of completed developments, the following examples of UK schemes cover rural settings, urban areas and town centres.

- Laughton Lodge, the Community Project (www.laughtonlodge.org), is a non-profit company limited by guarantee, with 21 houses and sharing 23 acres of meadowland, created on a redundant rural hospital site in

Sussex. The community centre building comprises offices, hall, training room, sun lounge, music room, games and playrooms, dining and cooking facilities and guest rooms.

- The Threshold Centre (www.thresholdcentre.org.uk), based at Cole Street Farm in Dorset, consists of 14 dwellings, half of which are occupied for rent and shared ownership in partnership with a local housing association. There is a common house with shared facilities, guest rooms, 'green energy' systems and a community market garden.
- LILAC: Low Impact Living Affordably Community (www.lilac.coop) is a cohousing community of 20 eco-build households in West Leeds. The homes and land are managed by residents through a pioneering financial model for mutual homeownership that links household payments to incomes and is structured to ensure permanent affordability.
- Springhill Cohousing (www.therightplace.net/coco/public) is near the centre of Stroud in Gloucestershire. It was the first new-build cohousing scheme to be completed in the UK. There are 34 units, ranging from one-bedroomed flats to five-bedroomed houses, and a three-storey common house and kitchen where meals are cooked and served during the week.
- Forgebank/Lancaster Cohousing (www.lancastercohousing.org.uk) is on the edge of Halton village just outside Lancaster city, with around 65 adults and 15 children resident. Homes were built to Passivhaus standards and have use of renewable technologies from solar, biomass and hydroelectricity sources. There is a common house for shared meals, guest rooms, co-operative food store and access to using other shared resources, including adjacent community workspaces.
- New Ground (www.owch.org.uk/structure-of-owch) is a development in Barnet, Greater London, created by the Older Women's Cohousing (OWCH) group, whose members range in age 'from the mid-fifties to around eighty'. The complex has 25 self-contained flats with shared communal facilities and gardens, comprising of 11 one-bed, 11 two-bed and 3 three-bed flats, plus a common room,

guest room, laundry and gardens. Seventeen flats are owned by their occupants on 250-year leases; eight are for social renters on assured tenancies and are managed by OWCH and Housing for Women.

- K1 Cohousing (www.cambridge-k1.co.uk) has been designed at Marmalade Lane in Cambridge, with 42 properties ranging in size from one-bedroom apartments to four-bedroom houses, some to be customised by the residents. There is a common house near the centre of the site, with car-free communal and garden areas placed between terraces of properties.
- Cannock Mill Cohousing (http://cannockmillcohousing colchester.co.uk/) is a development of new flats and houses in Colchester for members aged 40+, designed to Passivhaus and Lifetime Homes standards. The ancient mill building houses the scheme's communal and shared facilities.
- East Whins Cohousing 'cluster' is sited within Findhorn's eco-settlement land (https://4allsentientbeings.wordpress.com/2015/08/08/east-whins-cohousing-cluster-at-findhorn/), and is a custom-build approach for 25 residential units and modest shared facilities, developed by a private community-based housing design and development body.

### Policy implications for cohousing neighbourhoods

- The confusion introduced by using the umbrella/academic term 'co-housing' to denote highly divergent collaborative housing projects mars a proper appreciation of the distinctive contribution that *cohousing* settings offer for creating unique resident-led and -managed neighbourhoods.
- More practical examples are needed of how to support the economics of mixed-tenure cohousing settings, not least because a misguided interpretation of aspirant cohousing groups is that they are solely seeking dwellings to own. This does not reflect the range of households wanting a cohousing lifestyle but is a reflection of how difficult it has been to enable lower-resourced households to be housed within some final occupations.

- Some new UK cohousing projects have planned for greater numbers of residents than experience has suggested can support the relaxed and sustainable neighbourhood dynamics that a project's initiators have sought. The success of the intimacy and familiarity generated in neighbourhood schemes abroad has been based on a careful consideration of the interpersonal dynamics likely to emerge between certain numbers of households. Plans for UK cohousing schemes that do not recognise this experience will risk negating the grounds on which such neighbourhood intimacy could be found.
- There could be a temptation to consider that the innovative demands of designing a new cohousing neighbourhood's built environment is of greater importance than thinking about designs for its social environment, and that UK bodies with a 'development' expertise thereby come to believe they can adequately design cohousing opportunities in advance of prospective households coming together. 'Developer-led' cohousing neighbourhoods will be a contradiction in terms whenever crucial design and economic decisions lie outside of the resident members' amendment and control.

## Key organisations (cohousing)

- The UK Cohousing Network (https://cohousing.org.uk/) was established in 2007 to promote awareness of cohousing and support the development of new cohousing communities, and to act as a resource to enable members to share information, advertise related services and connect in order to discuss shared issues. In 2013 the Network established the 'UK Cohousing Trust' that now has charitable objectives for public benefit.
- The Cohousing Association of the United States (Coho/US; www.cohousing.org) is a national non-profit raising awareness of the benefits of cohousing and supporting the development of cohousing communities across the US.
- Bofællesskab.dk (http://xn--bofllesskab-c9a.dk) seeks to unite all community collectives in Denmark. The website aims to create a complete list of residential communities and collectives across the country.

- Landelijke Vereniging Centraal Wonen (National Association of Central Living; www.lvcw.nl/), based in Utrecht, is the central coordinating body for cohousing in the Netherlands.
- The Cohousing Company (www.cohousingco.com ) is the website of McCamant & Durrett Architects, based in America.

## Further reading

There is an extensive body of publications and papers available through the internet (see the following references), although the most extensive information in English has tended to draw from experience and analysis of US practice. The literature on specific cohousing practice from the UK is less extensive, although there are regular newspaper and magazine articles about the 'latest' cohousing initiatives, often used to draw attention to inviting interest in group living schemes.

Brenton, M. (1998), *We're in Charge: Cohousing Communities of Older People in the Netherlands*, Policy Press, Bristol, UK.

Brenton, M. (ed) (2008), *The Cohousing Approach to Lifetime Neighbourhoods*, Housing Learning & Improvement Network, London.

Brenton, M. (2013), *Senior Cohousing Communities: An Alternative Approach for the UK?*, JRF Publications, York, UK.

Bunker, S., Coates, C., Field, M. & How, J. (2011), *Cohousing in Britain: A Diggers & Dreamers Review*, Edge of Time Publications, BCM Edge, London.

Cojan, H.C. (2016), 'Factors that Influence the Continuity and Cohesiveness of North-West European Cohousing Communities', PhD dissertation, Cardiff Metropolitan University, Wales.

Durrett, C. (2009), *The Senior Cohousing Handbook*, New Society Publishers, Gabriola Island, BC, Canada.

Field, M. (2004), *Thinking about Cohousing: The Creation of Intentional Neighbourhoods*, Edge of Time Publications, BCM Edge, London.

Gehl, J. (1980/1987), *Life Between Buildings: Using Public Space*, Van Nostrand Reinhold, New York.

Hanson, C. (1996), *The Cohousing Handbook: Building a Place for Community*, Hartley & Marks, Vancouver, Canada.

Housing LIN (2017), 'Case Study 139: Community Building for Old Age: Breaking New Ground', Housing Learning & Improvement Network, London.

id22 (2012), *Cohousing Cultures: Handbook for Self-Organised, Community-Orientated and Sustainable Housing*, jovis Verlag, Berlin.

Jarvis, H. (2015), 'Towards a Deeper Understanding of the Social Architecture of Co-housing: Evidence from the UK, the USA and Australia', *Urban Research and Practice*, 8(1): 93–105.

Kärnekull, K. (2010), 'Cohousing in Sweden in the Second Half of Life', Cohousing Unit Färdknäppen: Collaborative Housing Conference, Stockholm, 5–9 May.

Kiesel, Y. (2018), 'Life in De Kersentuin: Examining the characteristics of a sustainable cohousing project', MSc Thesis, School of Social Sciences, University of Amsterdam.

Kim, G.H. (2006), 'Designing the Cohousing Common House', www.schemataworkshop.com.

McCamant, K. & Durrett, C. (1988, 1994), *Cohousing: A Contemporary Approach to Housing Ourselves*, Ten Speed Press, Berkeley, California, US.

McCamant, K. & Durrett, C. (2011), *Creating Cohousing: Building Sustainable Communities*, New Society Publishers, Gabriola Island, BC, Canada.

Navne, A.D. (1987), *At Bo i Fællesskab*, Artitektens Forlag, Copenhagen, Denmark.

Sargisson, L. (2010), 'Cohousing: A Utopian Property Alternative?', Centre for the Study of Social and Global Justice, University of Nottingham, UK.

Tummers, I.L. (2015), 'Introduction to the Special Issue: Towards a Long-term Perspective of Self-managed Collaborative Housing Initiatives', *Urban Research and Practice*, 8, *Special Issue* on 'Taking Apart Co-housing'.

Vestbro, D. (ed) (2010), *Living Together: Cohousing Ideas and Realities around the World,* Proceedings from Collaborative Housing Conference, Stockholm, 5–9 May.

## 2.6 ECO-HOUSING, 'LOW-IMPACT' SETTLEMENTS AND ECOVILLAGES

The desire to design and build dwellings that will support and protect social and natural environments has gone hand in hand with many forms of individual and communal projects. It has also steered plans and aspirations for larger and larger scales of community-led residence and settlement.

### Clarification of terms

- An eco-house is an environmentally low-impact home designed and built using materials and technology that reduces its carbon footprint and lowers its energy needs. The consistent approach is for this to result in lower environmental impacts throughout the whole lifecycle of a house – from site selection through design, choice of materials, construction, operation, maintenance and demolition.
- Eco-housing has been the term given to dwellings 'that have a low-impact on the environment' (Living Village Trust).
- Low impact has been defined as 'development that either enhances or does not significantly diminish environmental quality' (Diggers & Dreamers); it is commonly used to denote development which by virtue of its low environmental impact may be allowed in locations where conventional development is not permitted.
- An ecovillage is an intentional, traditional or urban community that is being consciously designed and created through locally owned, participatory processes. These have used four dimensions of sustainability – social, cultural, ecological and economic – to shape social and natural environments. 'Ecovillages are living laboratories pioneering beautiful alternatives and innovative solutions. They are rural or urban settlements with vibrant social structures, vastly diverse, yet united in their actions towards low-impact, high-quality lifestyles'. (Global Network).
- Off-grid has become the term to denote lifestyles that do not involve or require the use of mainstream sources of energy or other utilities.
- The terms carbon footprint/carbon emissions refer to the impact that carbon emissions and use of other gases will have upon climate change in the atmosphere and across the planet.
- Green architecture/green design are terms that draw from a philosophy of architecture and design that advocates the prioritisation and conservation of sustainable energy sources, the reuse and safety of

building materials, and the siting of buildings with consideration of their impact on the wider environment.

---

## Key appeal (eco-housing)

It is clear from many personal testimonies and from professional advice that increasing care should be given regarding what impacts on or damages the natural ecological systems on which we depend for survival, and that damage must be minimised. In the case of buildings, that concerns the gaseous emissions of $CO_2$ and other pollutants from the energy and construction processes involved in their creation and use, and to reducing any toxification and destruction of natural habitats resulting from resource extraction, transport and subsequent factory treatment.

The construction of buildings is responsible for around 10% of overall energy use and carbon emissions in the UK, whilst the use of those buildings is responsible for around 50% of energy use – the majority of which is for heating. A low-impact building will reduce this use and the prospects of such damage, through the selection of materials used in its construction, and through careful design of how the use of energy and water resources required across its lifetime could be minimised.

Consideration of the kinds of buildings that could be created by a focus on natural materials and 'low-impact' designs and techniques, have included:[19]

- building and working with earth – 'cob' building, earth-sheltered houses, rammed-earth buildings, subterranean dwellings;
- building with stone materials – stone buildings, stone slating, lime products;
- building with timber – post and beam construction, timber framing, round-wood framing and construction;
- building with crops-based materials – straw-bale building, hemp building;

- building with recycled materials (in conjunction with the above) – tyres, bottles, construction 'waste';
- low-impact retrofitting and adaptations of existing buildings.

A desire to live and undertake day-to-day activities alongside like-minded people sharing such ecological concerns informs the off-grid developments that have come from small-scale ecological groups and the ambitions for establishing larger eco-settlements and ecovillages. At their core they all recognise the wisdom of E.F. Shumacher's seminal work 'Small is Beautiful (a Study of Economics as if People Mattered)',[20] that the positive attention and value we give to each element of our surroundings will materialise in the qualities that we are storing up for our future. A protective regard for the natural environment can, however, practically coexist alongside other contemporary demands for work, leisure and commerce.

## Recent context (eco-housing)

Over decades, eco-housing activists have incorporated a wide variety of social and technological innovations into how their housing can adapt to given sites, and research depositories, like Grassroots Innovations,[21] offer a myriad of projects, kits and examples to explore. Attempts to create broader new ecovillages or eco-projects in the UK have, however, met with mixed success,[22] and projects have mainly sought to establish themselves in rural areas and take advantage of their natural environments, although some commentators have criticised initiatives that bring people into more isolated areas.) The BedZED mixed-use community,[23] completed in 2002, with 100 homes, office space, a college and community facilities, was a pioneering approach to create a model for an urban ecovillage, yet duplications on this scale have yet to occur.

Some high-level political proposals have sought to promote opportunities for 'sustainable living' on a large scale. The Sustainable Communities Plan in 2003, although intended to stimulate extensive amounts of new housing across the middle and southern parts of England (and removals of older properties in northern areas), also listed what was deemed to be characteristics of 'sustainable' places.[24] The Eco Towns Programme from 2007 subsequently gave an even stronger focus to some of those ideas,[25] promoting a balance between the provision of high-quality eco-standards for the built environment alongside similarly ambitious standards for wider social and economic environments. Both programmes ultimately failed to have a wider influence, not least due to the scale of local resistance that became evident, but some of the original eco-town proposals continue to have local impact, such as at Bicester and at Whitehill.

## Examples (eco-housing)

- The best-known UK example of a community operating in support of explicit ecological values remains the 'intentional community' at Findhorn in Scotland (www.findhorn.org). This is still expanding after its initial creation over 50 years ago and is internationally recognised for its engagement in support of 'sustainability' and sustainable living. It includes ecologically designed houses, innovative use of local building materials and neighbourhood technologies like a recycled sewage-treatment facility and electricity-generating wind turbines.
- Small groups based on rural smallholdings include the Lammas Community (http://lammas.org.uk) and Brithdir Mawr (www.brithdirmawr.co.uk), both in Pembrokeshire. These are off-grid and combine rural residences with recycling and conserving resources and farming the land according to organic principles.
- Hockerton Housing Project (www.hockertonhousingproject.org.uk/) is an exemplary community business and self-sufficient eco-housing development. Its earth-sheltered homes are built with high thermal mass and high levels of insulation to eliminate the need for heating systems. Residents of the five houses generate most of their own

energy, grow most of their own food, harvest their own water and recycle waste materials.

- Tinkers Bubble (www.tinkersbubble.org/) is a small woodland community which uses environmentally sound methods of working the land without using fossil fuels. It has planning permission for self-built houses on the condition that it makes a living from the land – mainly through forestry, orchards and gardening. As a result, it considers itself 'money poor but otherwise rich!'
- LILAC (Low Impact Living Affordable Community; www.lilac.coop) focused the design of its dwellings to incorporate straw-bale wall construction between lightweight timber structures, coated with lime render.
- The Ecological Land Co-operative (ELC; http://ecologicalland.coop) was set up to address the lack of affordable sites for ecological land-based businesses in England and Wales. It supports rural regeneration by developing affordable sites for farming, forestry and other rural enterprises which are viable and ecologically beneficial.
- The Living Village Trust (www.thelivingvillagetrust.com) was set up in 1993 'with a philosophy that embraces the concept of building sustainable communities where people genuinely want to live and interact with one another to enjoy a high quality of life'. It has designed and built a series of eco-friendly neighbourhoods to complement rural and urban growth around the UK.
- Land Matters (http://landmatters.org.uk/) is a rural permaculture project in Devon working to promote land-based communal living. Its aim is for its members to live by 'low-impact solutions that reduce our ecological footprint, so that we can be an active part of creating a sustainable world'.
- Lowland crofting, the principle 'of living on and working a small agricultural holding (Scottish Crofting Federation) is receiving a contemporary stimulus to create new woodlands, low-density housing, business and farms across central and lowland Scotland.

## Policy implications for eco-housing and eco-settlements

- The 'One World' approach in Wales, and the 'community empowerment' potential being established in Scotland, offer models for what regulatory systems might permit elsewhere in the UK, particularly in England. Care is still required, however, that their application does not relegate eco-solutions to remote places alone, as if there is no relevance to promoting sustainability in mainstream settlements.
- Strong interest has developed in 'passive house' design as a standard for dwellings to be energy-efficient, comfortable, affordable and ecological, all at the same time. By designing the technical aspects of efficient and insulated buildings to use only the energy sources inside the building, such as body or cooking heat from resident activities, or solar heat entering the building, such dwellings can substantially reduce the emission of greenhouse gases.
- The wider debate concerning 'what makes a community sustainable?', is more than a debate over the nature of what (eco-)buildings it may use. In one sense, all the examples in this publication represent solutions to make individual and collective lives 'sustainable', albeit each may emphasise diverse priorities. The principles of eco-housing and eco-settlements can clearly be incorporated within many approaches to designing the places and means in which people can wish to live.
- The Off Grid Festival (http://offgrid-festival.co.uk/about/) is a yearly exploration of community living, creativity and consciousness, sharing skills, trade knowledge and 'imparting wisdom', set up by the Association of Resilient Living. It is an experimental innovation lab for a new paradigm of economic, ecological and community thought, gathering people together who are making a difference in their lives and emergent cultures.

## Key organisations (eco-housing)

- Global Ecovillage Network (GEN; https://ecovillage.org) is 'a growing network of regenerative communities and initiatives that bridge cultures, countries, and continents … building bridges between policymakers, governments, NGOs, academics, entrepreneurs, activists, community networks and ecologically minded individuals to develop strategies for a global transition to resilient communities and cultures'. The network is made up of approximately 10,000 communities and related projects where people are intent on living together in greater ecological harmony with each other and their surroundings.

- Baltic Ecovillage Network (BEN; http://balticecovillages.net/home/index.htm) is an national association of ecovillage networks promoting environmental protection and restoration of nature through education, social methods and technical solutions, youth work, culture, intercultural understanding between people and environmental protection. The vision of the Association is to work through ecovillages in order to integrate human activities and the use of environmentally friendly lifestyles to promote the holistic development of human society.
- The Permaculture Association (www.permaculture.org.uk). Permaculture is a design process that helps 'design intelligent systems which meet human needs whilst enhancing biodiversity, reducing our impact on the planet, and creating a fairer world' for all. The Association supports members and others to design thriving communities across Britain by applying permaculture ethics and principles, and to contribute to permaculture worldwide.
- Association of Environmental Conscious Building: AECB (www.aecb.net) is a network of individuals and companies with a common aim of promoting sustainable building. It brings together builders, architects, designers, manufacturers, students, educators, housing associations and local authorities to develop, share and promote best practice in environmentally sustainable building.
- Centre for Alternative Technology (CAT; www.cat.org.uk) is an education and visitor centre demonstrating practical solutions for sustainability, covering all aspects of green living: environmental building, eco-sanitation, woodland management, renewable energy, energy efficiency and organic growing. Its base at Machynlleth in mid-Wales is a unique and valuable practical demonstration site, with an

enormous range of live examples of sustainable solutions and renewable systems, a visitor centre, graduate school and education centre.

- Builders Without Borders (www.builderswithoutborders.org) is an international network of ecological builders who advocate the use of straw, earth and other local, affordable materials in construction. To this end it organises and promotes hands-on workshops and learning events, and creates and partners with other organizations to build affordable sustainable housing.
- The Land Is Ours (http://tlio.org.uk/) campaigns peacefully for access to the land, to its resources and to the decision-making processes affecting them. It has three main aims: 'Land for homes; Land for Livelihoods; and Land for Life'.
- Chapter 7 (http://tlio.org.uk/chapter7/) – the 'planning office' of The Land Is Ours – gives planning advice to people seeking to/or already embarking upon living on the land, or engaged in land-based livelihoods.
- The Scottish Crofting Federation (www.crofting.org/faqs/67) was established by crofters themselves and is dedicated to campaigning for crofters and for the future of crofting. It claims to be the largest association of small-scale food producers in the UK.
- The Transition Network (https://transitionnetwork.org/) 'is a movement of communities coming together to reimagine and rebuild our world'. It has been expanding since 2005 to support local groups and communities in how they 'step up to address the big challenges they face by starting local'. There are a number of 'Transition Towns' in the UK campaigning for holistic improvements across a wide range of issues and facilities.
- Social Farms and Gardens (www.farmgarden.org.uk/) is a recent amalgamation of long-standing national organisations that help grassroots groups create the community farms and gardens, school farms, care farms, wildlife and roof gardens, community orchards, community-run allotments

and community-supported agriculture schemes that complement aspirations to foster sustainable community housing spaces.

## Further reading

Bang, J.M. (2005), *Ecovillages: A Practical Guide to Sustainable Communities,* Floris Books, Edinburgh, Scotland.

Bird, C. (2010), *Local Sustainable Homes: How to Make Them Happen in Your Community*, Transition Books, Totnes, Devon, UK.

Bunker, S., Coates, C., Dennis, J. & How, J. (2014), *Low Impact Living Communities in Britain: A Diggers & Dreamers Review*, Edge of Time Publications, BCM Edge, London.

Centre for Alternative Technology (2017), *Zero Carbon Britain: Making It Happen*, CAT, Machynlleth, Wales.

Chatterton, P. (2014), *Low Impact Living: A Field Guide to Ecological, Affordable Community Building*, Earthscan Publications, London.

Department of Communities and Local Government (2007), *Eco-towns Prospectus,* HM Stationery Office, London.

Eurotopia (2014), 'The Eurotopia Directory of Communities and Ecovillages in Europe', www.eurotopia.de.

Fairlie, S. (1996), *Low Impact Development: Planning and People in a Sustainable Countryside,* Jon Carpenter Publishing, Charllbury, UK.

Jackson, H. (1998), 'What Is an Ecovillage?', paper presented at the Gaia Trust Education Seminar, Thy, Denmark, September.

Jackson, H. (2004), 'The Ecovillage Movement', *Permaculture,* 40: 25–30.

Lewinsohn, S. (2008), 'Planning Policy and Low Impact Developments', MSc Thesis, Centre for Alternative Technology & University of East London.

Mann, J. (2008), 'Eco-towns, New Labour and Sustainable Development', *People, Place & Policy*, 2(3): 132–9.

Martel, T. (2018), 'Effective Plans for Our Future', www.aecb.net/effective-plans-for-our-future/.

National Assembly for Wales (2009), 'One Wales One Planet', www.assembly.wales/gen-ld8219-e.pdf.

Pickerill, J. (2016), *Eco-Homes: People, Place and Politics (Just Sustainabilities),* Zed Books, London.

Seyfang, G. (2010), 'Community Action for Sustainable Housing: Building a Low-carbon Future', *Energy Policy*, 38: 7624–33.

University of the West of England/Land Use Consultants (2002), 'Low Impact Development: Planning Policy and Practice, Final Report', University of the West of England, Bristol, UK.

Woolley, T. (2013), *Low Impact Building: Housing Using Renewable Materials*, Wiley, Chichester, UK.

## 2.7 'INTENTIONAL COMMUNITIES' AND LIVING IN GROUPS

Desires for living communal lifestyles alongside others holding common social, political, or spiritual values continue to be a strong motivation into the 21st century, albeit these may be below many people's radar.

---

### Clarification of terms

- An intentional community (IC) is 'a group of people living together with some shared resources, on the basis of explicit common values'. (Fellowship for Intentional Communities).
- The term community is used here for 'when a group of people live together intentionally in order to pursue ideological (and material) goals together ... when people care to abandon a little of their individual privacy in order to communally create personal aspects of life' (eurotopia – see list below).
- Commune, originally a word to denote the smallest administrative division in countries like France, Italy and Switzerland, is here used to define 'a group of people, drawn from more than one family or kinship group, who have voluntarily come together to share certain aspects of their lives together, and who are characterised by a certain consciousness of themselves as a group' (Communes in Britain).
- A utopian community has been classified as one 'trying to make the world a better place for more than just individuals, which might be for its members or for society as a whole' (Utopia Britannica – see below).
- An alternative community has been defined as one 'distinguished from so-called 'straight' society mainly by economic and relational arrangements' (Directory of Alternative Communities).

- A gated community is a description given to different kinds of 'walled or fenced housing development, to which public access is restricted' (academia.eu).

## Key appeal (intentional communities)

Undertaking activity 'in common' with others is a persistent theme throughout this book. Wanting to connect closely with others is a strong motivation for people to arrange to be 'living in common' with 'like-minded' souls. That is easily evidenced by a whole range of communal living arrangements, from having one or more rooms in a shared house (or even mansion), with a shared kitchen, shared washing facilities, a rota for cooking and other monthly chores, through to diverse forms of separate self-contained accommodation in shared neighbourhoods or similar locations. It has involved some groups sharing incomes and possessions, and others sharing relationships. Many shared ideals are focused on how to live in communally sustainable and low-carbon lifestyles.

The descriptive use of the term intentional community has been in existence for years, with a typical use evoking ideas of groups associated with the 'hippy' culture of the 1960s and 1970s, likely to have an *anti*-establishment ethos. Whilst the term is still used to denote groups with that history, it has also been attached to examples of housing co-operatives, cohousing schemes, ecovillages or other low-impact groups – settings which can demonstrate elements of shared intentionality behind their core frameworks. Several complementary values are apparent in what the members of communal settings may seek:

### *Communities that maximise a focus upon a social life*

Despite a general sense that 'we are all social animals', it is fair to point to an uncertainty in the UK about whether modern people will be truly at ease living closely connected to others, particularly to neighbours. Adulthood is associated with being autonomous, whereas individuals living in 'shared' communities must have to accept a loss of personal control that will constrain such autonomy. A key figure behind the UK's first new-build cohousing scheme in Stroud described to this writer his feeling that the British see living in a 'shared space' as tantamount to a backwards step in society, that what people 'really want' is to have their own car next to their own house, with their own front garden – always with aspirations to move into yet bigger and more separate houses in the future. A 2015 report from Demos[26] on whether community-led schemes might challenge contemporary UK housebuilding, included some idiosyncratic collection of remarks from potential neighbours in village settings about the ideas and values of new 'community' schemes. One observation quoted a sense of discomfort in the idea of 'eating in common' with one's neighbours – as if it was not the 'done thing'.

People choosing to live a communal life with others challenge any such uncertainty about social connectedness, pointing to a clear sense of social well-being in living as part of a such a group, different from an identity as part of a 'nuclear' family. They point to a core feeling that communal settings are experienced as invariably 'safe, [where] it is great to know your neighbours and there is always good, cheap food around the table'.

### *Communities that uphold strong political values*

The clear political intention behind the establishment of some UK communal settings has been ably described in the histories of communist/socialist and back-to-the-land settlements (see publications from Darley, Hardy, Pearson and Ward below). These often note the influence of 'communitarianism', the philosophy that emphasises the positive influence on personal identity exerted by and through wider community relationships.

Shared political values have certainly informed places in which 'co-operation' with others is prized, as evident in the income-sharing and economic ideals of diverse residential groups, and in the activities of squatters' groups. Under the broad banner of 'green' politics it has also driven ecological or low-impact approaches, along with an overt focus by some communities on forms of diet, or child-rearing, or use of renewables, or using particular building materials.

#### *Communities that have a shared religious life, or a 'faith-friendly' environment*

Living with others of a similar religious view on life may not usually be included with references to 'intentional communities', yet such communities exert a perennial attraction for people seeking a particular context for a 'communal' life. The UK has retained a probably under-appreciated range of traditional Christian monasteries and convents,[27] besides other long-standing faith-based communities like the Bruderof communities and a growing number of 'new monastic' settlements, such as the Northumbria Community.[28] An increasing number of new residential communities are also emerging, in urban and rural settings, that adhere to faiths other than Christianity – such as Buddhist, Zen and Hindu traditions – many of which offer a chance to visit or to stay.[29]

### Recent context (intentional communities)

How many people are now living within 'intentional communities' in the UK? Notwithstanding any self-selective approach that characterises the range of communal groups known to the Diggers & Dreamers Collective, in 2009 the editors of the D&D Directory provided a 'guesstimate of this number being in the region of 23,000'[30] – based on the average size of the known and verifiable groups in existence over the past 25 years, multiplied by a factor of 5 for groups still 'not known'. (This may not be a scientific assessment, but it offers a number that could be challenged more methodically, if others so wish.)

Setting up new 'shared' neighbourhoods and villages continues to motivate cohousing and ecovillage schemes, and new co-operatives continue to be established in houses where residents share some or all of the facilities.

## Examples (intentional communities)

- The Redfield Community in Buckinghamshire describes itself as 'an Intentional Community … with a legal structure of a Fully Mutual Housing Co-operative'.[31] Around 20 adults and children live in a large mansion with 17 acres of gardens, woodland and pasture, surrounded by farmland. No rooms are permanently occupied by the same members, but their use is continually reviewed to meet changing personal needs. The on-site Redfield Centre offers self-catering and catered accommodation for other educational and leisure purposes.
- The Rainbow Housing Co-operative took over the management of an entire street in the 1970s as a tenant-run co-operative, and then in 1992 bought it from Milton Keynes Development Corporation under the Right to Buy scheme.[32]
- Cannon Frome Court (www.canonfromecourt.org.uk) is a 'farming co-operative and intentional community in the Herefordshire countryside', on a 40-acre organic farm, with about 50 (adults and children) living in a Georgian manor and stable-block divided into 19 self-contained homes. The land and buildings are owned by Windflower Housing Association, a 'co-ownership' association, and homes are purchased on a 999-year lease.
- Whiteway Colony is a Tolstoyan community set up 120 years ago by socialists to be a pioneering residential enterprise. Today it occupies the same 40-acre site, where colonists live in their own individual homes on communally held land.[33]
- The Lammas project in South Wales (http://lammas.org.uk), centred around the ecovillage of Tir y Gafel in North Pembrokeshire, 'has been designed as a model that can be replicated … to pioneer living on the land. It empowers

people to explore what it is to live a one-planet lifestyle [and] that alternatives are possible'. It consists of nine smallholdings, a range of peripheral projects and networks, and a 'Community Hub' building at the centre of the site, which hosts educational and recreational activities.

- Findhorn (www.findhorn.org) is the communal setting in the UK with probably the most prominent international profile and reputation. It has a community of more than 500 people who endeavour 'to live the vision of creating a better world by starting with themselves'. It is visited by 'thousands of fellow-adventurers from all over the world' who participate in workshops, conferences and other special events focused on communal life and holistic growth. The Findhorn Foundation is co-founder of the Global Ecovillage Network and Holistic Centres Network.
- BBC4 programmes in its 'Quiet TV' season of 2017 depicted three Benedictine Christian monastic communities from England and Scotland – Downside Abbey, Plusgarden Abbey and Belmont Abbey – ranging in size from 18 to 40 monks. Throssel Hole (http://throssel.org.uk) is a Zen Buddhist monastery of lay and ordained members located in Northumberland, affiliated to the Order of Buddhist Contemplatives (https://obcon.org). Samye Ling (www.samyeling.org) is a large Tibetan Buddhist temple and monastic community of over 60 people, located in a quiet rural setting inside the southern border of Scotland.

Other examples of shared 'supportive' settings in the UK include Camphill Communities (www.camphill.org.uk) and L'Arche (www.larche.org.uk) for people 'with learning disabilities, mental health problems and other special needs', and the Pilsdon Community (www.pilsdon.org.uk), founded principally to help people in crisis develop coping skills for their future life and work. These may seem at face value distinct from the core focus on the community settings already noted above, although long-term personnel resident in some

Camphill communities are investigating how to establish group living schemes for their own retirement.

Communities in the UK do accept visitors, when arranged appropriately, and some open themselves for sessions and weekends when they arrange structured visits. One link could be through a 'working weekend', such as those organised in conjunction with World Wide Opportunities on Organic Farms (www.wwoof.org.uk), and Social Farms and Gardens (www.farmgarden.org.uk/about-us), whereby volunteering can be undertaken in exchange for food, accommodation and the opportunity to learn more about a host community. A key annual source of information on communities and for visits or 'retreats' is the yearly handbook issued by the Retreat Association (www.retreats.org.uk/retreatshandbook.php).

### Policy implications for 'intentional' communities

- The sense in which contemporary UK housing delivery is set up to provide separate dwellings, somewhat disconnected from their surroundings, will not satisfy the aspirations of those wanting more collective ways to live. Yet mainstream housing and planning delivery practice can also benefit from helping proposals for new 'intentional communities' to populate parts of larger new-build areas – for example, US developers have deliberately used cohousing schemes as providers of high-quality collaborative neighbourhoods to attract other consumers to their broader real estate developments.
- The variety of contemporary forms that could be used by 'intentional' groups are well suited for inclusion in plans for the redevelopment or regeneration of current urban settings. This is particularly so with groups from households with common cultural ties that could establish group living provisions for local households from a similar or shared background.
- The flexibility of the 'One Planet' planning policy in Wales is a positive example of how wider productive policy can evolve from work to meet the aspirations originating from parochial interests. In this case it has enabled groups to live in more ecological and communal settings, being as close to 'off-grid' as possible. It is also demonstrably something with very contemporary interest – there are new groups keen to make use of shared living arrangements to maximise climate support initiatives.

## Key organisations (intentional communities)

- Diggers & Dreamers (www.diggersanddreamers.org.uk) is a not-for-profit collective whose aim is 'to dispel the myth that communes came and went with the 1960's, and to bring the idea of communal living to the attention of more people'. The collective publishes articles, books and directories (publishing is through Edge of Time – www.edgeoftime.co.uk) and keeps a website of intentional communities and communes, 'where people can find out about living communally in the UK'.
- Eurotopia: Living in Community (www.eurotopiaversand.de) is a network of intentional communities across the globe, and the producers of an international directory of many different kinds of communal groups and residential settings.
- The Fellowship for Intentional Community (FIC; https://www.ic.org) is a non-profit organisation dedicated to supporting and promoting the development of intentional communities (which, it notes, may include ecovillages, cohousing, residential land trusts, communes and housing co-operatives), and the evolution of co-operative culture. It operates an international directory of communities and communal settings and produces a magazine, *Communities*, four times a year to disseminate information, stories and ideas about intentional communities.
- The International Communal Studies Association (www.communa.org.il/icsa) was formed in 1985 to provide a common framework for a scholarly exchange of information regarding 'communal life, communes, intentional communities, kibbutzim and other collective communities throughout the world'. International conferences organised at three-year intervals enable scholars and members of all kinds of intentional communities to meet and exchange views, experiences of communal life and other research.
- The Society for Utopian Studies (SUS; http://utopian-studies.org) is an international, interdisciplinary association

studying utopianism in all its forms, with a particular emphasis on literary and experimental utopias.

- The Federation of Egalitarian Communities (www.thefec.org) is a union of Egalitarian Communities – communities holding land, labour, income and other resources in common – which have joined together to promote 'a lifestyle based on Equality, Cooperation, and Harmony with the Earth'.
- Intentional Communities Research and Enterprise Group, Cardiff Metropolitan University (www.cardiffmet.ac.uk/health/intentionalcommunities) is a research group drawing academics and practitioners together to look at 'how Intentional Communities can contribute to the health and well-being of people'.

## Further reading

Blandy, S., Lister, S., Atkinson, R. & Flint, J. (2002), *Gated Communities: A Systematic Review of the Evidence*, Sheffield Hallam University/University of Glasgow, Glasgow, Scotland.

Cray, G., Mobsby, I. & Kennedy, A. (2010), *New Monasticism as Fresh Expression of Church*, Canterbury Press, Norwich, UK.

Darley, G. (1975/2007), *Villages of Vision: A Study of Strange Utopias*, Five Leaves Publications, Nottingham, UK.

Diggers & Dreamers, *The Guide to Communal Living* (annually through to 2008/9), www.diggersanddreamers.org.uk.

Dunstan, P. (2017), *Anglican Religious Life 2018–19: A Yearbook of Religious Orders and Communities in the Anglican Communion*, Canterbury Press, Norwich, UK.

Eurotopia: Living in Community (2016), *Directory*, www.eurotopiaversand.de

Gorman, C. (1971), *Making Communes*, Whole Earth Tools, London.

Hardy, D. (1979), *Alternative Communities in Nineteenth Century England*, Longman, London.

Hardy, D. & Ward, C. (2004), *Arcadia for All: The Legacy of a Makeshift Landscape*, Five Leaves Publications, Nottingham, UK.

Hinton, J. (1993), *Communities – Stories of Christian Communities in Europe*, Eagle Publications, Guildford, UK.

Jamison, C. (2006), *Finding Sanctuary: Monastic Steps for Everyday Life*, Weidenfeld & Nicholson, London.

Jones, T. (2007), *Utopian Dreams: In Search of a Good Life*, Faber and Faber, London.

Pearson, L.F. (1988), *The Architectural and Social History of Co-operative Living*, Macmillan Press, Basingstoke, Hampshire, UK.

The Retreat Association (2019), 'Retreats 2019', The Retreat Association, UK www.retreats.org.uk/retreatshandbook.php.

Rigby, A. (1974), *Communes in Britain*, Routledge & Kegan Paul, London.

Tawney, R.H. (1964), *The Radical Tradition*, Pelican Books, Harmondsworth, Middlesex, UK.

Vallance, E. (2009), *A Radical History of Britain*, Little Brown, London.

Ward, H. & Wild, J. (2006), *The Monastic Way*, Canterbury Press, Norwich, UK.

## 2.8 HOMES TO TRAVEL WITH

Not all 'homes' are made of bricks or timber or are fixed places of residence. Mobile dwellings, or other non-permanent arrangements, are still preferred by some households rather than being tied to a single place.

### Clarification of terms

- Gypsies and Travellers are described as 'Persons of nomadic habit of life, whatever their race or origin, including such persons who on grounds only of their own or their family's or dependants' educational or health needs or old age have ceased to travel temporarily' but excluding members of an organised group of travelling show people or circus people as such (Department of Communities & Local Government).

  Article 8 the European Convention of Human Rights (providing a right to respect private and family life), notes that 'occupation of [a] caravan is an integral part of [...] ethnic identity as a gypsy, reflecting the long tradition of that minority of following a travelling lifestyle ... [even though] many gypsies increasingly settle for long periods on one place' [European Convention of Human Rights].

The list of ethnically defined 'gypsy' groups includes English or Welsh Gypsies (together labelled as Romany Gypsies, protected from discrimination under the Equalities Act 2010), Scottish Gypsy Travellers, Irish Travellers and Roma.

- Gypsy and Traveller Accommodation Assessments are assessments to compile information about an area's local Gypsies and Travellers, which aims to generate reliable estimates of the need for future accommodation and provisions (Department of Communities & Local Government).
- New Age Travellers are 'a relatively recent phenomenon, linking back to the Hippy culture of the 1960's and 70's, and more a lifestyle choice than an ethnicity; most New Age Travellers were born into a "conventional" lifestyle and made a conscious decision to leave settled life and become a traveller, although there are now second and third generation travellers who have been born into travelling families'.[34]

  'New' or 'New Age' Travellers may be recognised by the diverse assortment of vehicles in which they live (vans, lorries, coaches, double-deckers, ex-ambulances and caravans converted into mobile homes), or use of other temporary accommodation, like bender tents, tipis and yurts.
- Bargee/bargee traveller/canal-dwellers are those who live on, and travel in, one or other kind of boat throughout the year and as a deliberate lifestyle.
- Liveaboards make a boat, small yacht or barge in a marina their primary residence, though it may be chosen for an integration of transportation, recreation and housing purposes.

## Key appeal (travelling homes)

Why do some people move their homes around? Because they can, and have, for many years.

There are ancient UK traditions of 'roaming' as a lifestyle that is distinct from living in fixed or mainstream accommodation, with age-old similarities to other distant cultures with extensive nomadic traditions. Travelling is judged to be less detrimental to the natural environment than building something solid and permanent, and it may not in any case be possible to find secure or affordable accommodation of choice in more 'settled' locations. All travellers, including New Age Travellers, have recently had a 'right to roam' protected by human rights legislation (supplemented by subsequent Acts and legal decisions concerning sites, evictions and discrimination).

A traveller's quote in the 2004 'Thirteenth Report of the Select Committee of the Office of the Deputy Prime Minister on England's Irish Travellers' aptly demonstrates the impulse to live a travelling life:

> when Travellers speak of Travelling, we mean something different from what country people [sedentary people] usually understand by it … the physical fact of moving is just one aspect of a nomadic mind-set that permeates every aspect of our lives. Nomadism entails a way of looking at the world, a different way of perceiving things, a different attitude to accommodation, to work, to life in general. (Thirteenth Report, Para 55)

Some travellers may admittedly stay on a site for an extended period of time whilst children attend local schools and vehicles are repaired, and it may be problematic for some households with limited resources to move from site to site at short notice.

A readiness to adopt a nomadic lifestyle also permeates the UK's New Age culture, with its readiness to travel for seasonal employments or to festivals. This 'new' culture draws in part from the nature of grassroots social and political resistance during the 1980s and 1990s. These included the Peace Camps set up outside UK military bases in protest against the presence of nuclear weapons in Europe, and other encampments set up to protest against major UK road constructions.

The appeal of a parallel lifestyle as residential boaters, or liveaboards, attracts households comfortable with the basic conditions of 'continuous cruising' – needing to move a boat from mooring to mooring officially every 14 days – but can be problematic when households wish to consolidate local jobs, or to access schools or other amenities. Living on canals or in marinas might seem less expensive (more 'affordable') than living on land, but there has been very little systematic research undertaken on the extent to which this is truly down to a choice or to necessity.

## Recent context (travelling homes)

How many people consider themselves to be 'travellers' in the UK? Who may live in the 'non-traditional' dwellings noted here? The 2011 Census included a new ethnic category to collect data on Gypsy, Traveller and Irish Traveller communities. This included detail of non-permanent or 'temporary' dwellings (such as caravans, mobile homes, converted railway carriages and houseboats) if they were the occupant's main residence and council tax is payable as a main residence. This identified around 63,000 people in the UK describing themselves as members of these groups (about 0.1% of the population), of which 58,000 were living in England and Wales. (The South East region of England had both the largest number of Gypsies and Irish Travellers and the largest number per 10,000 people.)

Other sources suggest, however, that the 2011 Census figures may be gross underestimates. The 'Traveller Movement organisation' believes a majority of informed estimates put the population at between 180,000 and 300,000 – a difference from official government statistics (based upon the bi-annual Caravan Count), partly due to diverse definitions and partly because 'many are believed to be living in actual houses'. One recent House of Commons Briefing Paper (08083, September 2017) expressed concern on a lack of robust data for overall numbers.

In determining whether persons are 'gypsies and travellers' for the purposes of UK planning policy, formal regulations now state that consideration should be given to:

- whether people previously led a nomadic habit of life;
- whether there is an intention of living a nomadic habit of life in the future; and if so
- how soon and in what circumstances.

Requirements are placed on all local planning authorities to identify sites or other provisions that could be suitable for meeting the needs of traveller communities (excluding groups travelling for the purposes of holding fairs, circuses or shows). Assessments are invariably for the numbers of new permanent 'pitches' for caravans, alongside other transit pitches. (Modern permanent pitches can include space for a static caravan and other vehicles, plus use of an amenity block with a kitchen, bathroom and dayroom.)

Formal assessments of where and how these could be met routinely attract strong local resistance. One consequence is the emergence of ad hoc encampments, often in very marginal and disruptive settings, such as wide verges or other overgrowth. Even when the members of some Gypsy and Traveller communities have purchased sites to take charge of their own planned solutions, there has been conflict with existing local residents and other stakeholders. At times when a purchased site does not meet official approval, this has even led to eviction from their own land.

In terms of assessing how many bargees or liveaboards are in the UK, the Residential Boat Owners Association (RBOA) estimated in 2015 that 'about 15,000 people are thought to live afloat in Great Britain' – an informed guess, based on an earlier 2005 estimate of 10,000–15,000 by the UK government (ODPM, 2005). This has historically been focused on numbers from only a few of the larger UK navigation authorities, not all of them. There are about 5,000 boats, including those with permanent or long-term moorings, within just the Canal and River Trust's London Waterways region. Over 10,000 people are believed to live on the capital's waterways.

The Housing & Planning Act 2016 did include some broad guidelines for local planning authorities on how to assess local demand for using boats and canal craft as places of 'primary residence'. Some draft central government guidance on assessing the 'Housing Need for Caravan and Boat Dwellers' was issued the same year, but little information is yet available on how this is being applied in practice.

Yet as the number of boats and watercraft being used on a rental basis increases, so too is the welfare system adapting to wider demands from non-traditional forms of accommodation. From June 2017 boat-owners with permanent moorings can apply for a 'Static Letting Licence' as a permit for long- and short-term boat rentals (some planning permissions may be required). Households renting such spaces could then claim housing benefit towards mooring fees and rent charges.

### Examples (travelling homes)

The Ministry of Housing, Communities and Local Government conducts twice-yearly counts in January and July of the number of caravans on authorised and unauthorised sites across England. The 'Count of Traveller caravans – January 2019' was issued in June 2019 on the respective numbers of travellers' caravans identified for each local authority in England (it did not cover static caravan accommodation, or other 'park homes'). The total number of traveller caravans in England in January 2019 was 22,662 – 284 fewer than the 22,946 reported in January 2018; 6,590 caravans were on authorised socially rented sites, a decrease of 334 since the January 2018 count.

The management of the 'pitches' on local sites is usually carried out by local authorities or by private organisations. Other provision is emerging from within the travellers' communities themselves: the 'Friends, Families and Travellers' network has identified projects in the Mendip area, in Brighton and in Bristol of Gypsy and Traveller groups turning to potential community land trust frameworks in order to establish their own sites. Mendip District Council has a small guidance leaflet to assist Gypsy and Traveller communities looking to identify land that might be suitable for sites.

Other independent bodies have been established to provide broad support to local Gypsy and Traveller communities, such as the 'TravellerSpace' charity (www.travellerspace-cornwall.org) supporting communities in Cornwall and the South West to have a voice in policy formation; Leeds Gate (http://leedsgate.co.uk/), a members-based body, working to support communities in West Yorkshire; and the London and Gypsy Travellers organisation (www.londongypsiesandtravellers.org.uk), set up to help communities 'gain more influence on the decisions that affect them'.

Information on examples of 'New Age' communities is less abundant than other Gypsy and Traveller groups. The 2012 publication *Travelling Daze* from Alan Dearling is a rare summary of New Age Travellers and festivals from the 1960s onwards. Websites exist describing encampments and converted vehicles in use and for sale.[35]

Examples of how boat-dwellers have been organising themselves can be evidenced in the groups leading different campaigns. Locally this can be like the Kennet and Avon Boating Community (https://kanda.boatingcommunity.org.uk/), active against limitations in the terms of mooring berths in the Bath and Avon area. A national 'Boats Are Homes' demonstration on Saturday 8 April 2017 was organised by the boating community to show resistance to the Canal and River Trust's implementation of mooring restrictions. It challenged the Trust's interpretation of required patterns of movement and its plans to clamp down on boaters with continuous cruiser licences who 'do not move often enough'.

### Policy implications for travellers' homes

Issues facing households that do not use 'permanent' dwellings as a fixed home are:

- Not all formal assessments of the needs of travellers are conducted as the guidance stipulates, and not all the completed assessments identify what future provision could or should emerge.
- Where the identification and location of sites for travellers is undertaken, they can be situated in environmentally deprived areas – on wasteland, floodplains or under motorway flyovers. This has a negative effect on the physical and mental well-being of travellers, compounded by a lack of access to other services suitable for travellers' needs.
- Travellers can also be formally considered as suitably housed when they have been allocated standard bricks-and-mortar accommodation.

This can mitigate against leading the nomadic lifestyle that is central to traveller culture and can also lead to neighbour disputes over differences between settled and nomadic lifestyles.

- Boat-dwellers consider that the right to use and live on a boat without a permanent mooring on Canals and Rivers Trust (CRT) waterways is enshrined in the British Waterways Act 1995, and that the CRT is acting unlawfully in imposing limitations based on a minimum distance of movement, or pattern of travel.
- More in-depth examination is required of the extent to which the pressures and costs of the broader UK housing market are driving households to live afloat, or to the use of associated amenities (like marinas) previously considered for temporary or leisure provision.

## Key organisations and information (travelling homes)

- Friends, Families and Travellers (www.gypsy-traveller.org/about-us/) is a national traveller-led charity which has been working for equality for Gypsies and Travellers since 1994, on behalf of all Gypsies, Roma and Travellers regardless of ethnicity, culture or background.
- The National Federation of Gypsy Liaison Groups (www.nationalgypsytravellerfederation.org) states its key aims as to promote social inclusion by working with Gypsy and Traveller groups that are socially excluded; provide a network group that encourages and enables members of the Gypsy and Traveller community to participate more effectively in the wider community; and to educate the public about the culture, life and traditions of the Gypsy and Traveller community.
- Gypsies and Travellers Wales (http://cgtp.co.uk/) aims to support Gypsies and Travellers in Wales to access suitable accommodation, public services and employment skills, and to encourage policies and practices which meet the needs of Gypsies and Traveller communities.
- The Traveller Movement (http://travellermovement.org.uk/) is a national 'civil society' organisation working in partnership with Irish Traveller, Gypsy and Roma communities challenging discrimination and promoting inclusion. It has a National Traveller Advisory Group made up of Travellers

and Gypsies from across the country as a key consultation point for the policy team on a wide range of issues, and produces a seasonal newsletter, 'Wheels in Motion'.

- The Travellers Times (http://travellerstimes.org.uk/About.aspx) provides the latest news, pictures, films, features and information from, and for, Gypsy, Roma and Traveller communities.
- The Canal and River Trust (https://canalrivertrust.org.uk/about-us) is the main national body (formerly known as British Waterways) overseeing the maintenance and operation of 2,000 miles of canals and rivers, including an enormous network of bridges, embankments, towpaths, aqueducts, docks and reservoirs. The CRT issues licences for canal use and moorings.
- Inland Waterways Association (IWA; www.waterways.org.uk/) is the leading membership charity dedicated to protecting and restoring Britain's network of canal and river waterways. It helps formulate national policies as well as undertake local and national campaigns. IWA is working with navigation authorities, waterway user representative bodies and boaters to address the issue of the limited number of moorings.
- The Residential Boat Owners' Association (RBOA; www.rboa.org.uk) was established in 1963. It is a national organisation representing and promoting the interests of people living on boats in the British Isles. It produces 'Soundings', a bi-monthly newsletter for members.
- The National Association of Boat Owners (http://nabo.org.uk/) was formed in 1991 to promote the interests of private boaters on Britain's canals, rivers and lakes, 'so that their voice can be heard when decisions are being made which might affect their boating'.
- The National Bargee Travellers Association (NBTA; www.bargee-traveller.org.uk) is a volunteer organisation formed in 2009 that campaigns and provides advice for itinerant boat-dwellers on Britain's inland and coastal waterways. This includes anyone whose home is a boat and who does

not have a permanent mooring for their boat with planning permission for residential use.

Other bodies that assist caravan and motorhome users – such as the Camping and Caravanning Club (www.campingandcaravanningclub.co.uk) and the Caravan and Motorhome Club (www.caravanclub.co.uk/discover/) – have historically been focused on leisure-based activities, rather than on the use of such vehicles as a primary residence. Similarly, the Barge Association (http://barges.org ) is predominantly for people interested in using barges for leisure purposes.

## Further reading


Brown, F. (1988), *Fairfield Folk: A History of the British Fairground and Its People*, Malvern Publishing, Upton upon Severn, UK.

Cadger, B. (2009), 'Gypsy/Traveller Numbers in the UK – A General Overview', BEMIS, http://bemis.org.uk/resources-info/information-for-gypsytravellers/reports/.

Corble, N. & Ford, A. (2017), *A Beginners Guide to Living on the Waterways*, The History Press, Stroud, UK.

Cromarty, H. (ed) (2017), 'Gypsies and Traveller', UK Parliament Briefing Paper 08083, House of Commons Library, London.

Dearling A. (2012), *Travelling Daze: Words and Images from the UK's New Travellers and Festivals*, (ed.) Enabler Publications, Eyemouth, Scotland.

Department of Communities and Local Government (2015), *Planning Policy for Traveller Sites*, DCLG, London.

Department of Communities and Local Government (2016), *Draft Guidance to Local Housing Authorities on the Periodical Review of Housing Needs – Caravans and Houseboats*, DCLG, London.

Field-Lewis, J. (2015), *My Cool Houseboat: An Inspirational Guide to Stylish Houseboats*, Pavilion, London.

Friends, Families & Travellers (2017), 'Briefing: Lack of Increase in Affordable Pitches for Gypsies and Travellers in England', FFT, www.gypsy-traveller.org/resource/lack-increase-affordable-pitches-gypsies-travellers-england/.

Irish Traveller Movement in Britain (2013), 'Gypsy and Traveller Population in England and the 2011 Census', www.travellermovement.org.uk.

Mendip District Council (2016), 'Guidance Leaflet for Gypsy & Traveller Site Location', MDC, UK.

Ministry of Housing, Communities and Local Government (2019), 'Count of Traveller Caravans, January 2019', HMSO, www.gov.uk/mhclg.

National Bargee Travellers Association (2017), 'Boat Dwellers Refuse to Be Kicked off the Waterways', *NDTA Newsletter*, May, https://nbtalondon.wordpress.com.

Opinion Research Services (2017), 'Leicester City and Leicestershire Gypsy, Traveller and Travelling Showpeople Accommodation Assessment', Swansea, UK.

Reeve, D. (1958), *Smoke in the Lanes*, Abacus Books, London.

Residential Boat Owners Association (2007), 'Living Afloat', www.rboa.org.uk.

Residential Boat Owners Association (2016), 'So You Want to Live on a Boat?', www.rboa.org.uk.

Richardson, J. & Ryder, A. (eds) (2012), *Gypsies and Travellers: Empowerment and Inclusion in British Society*, Policy Press, Bristol, UK.

Richardson, J. & Codona, J. (2016), *Managing and Delivering Gypsy and Traveller Sites: Negotiating Conflict*, CIH/JRF, Coventry, UK.

Ryder, A. (2012), 'Hearing the Voice of Gypsies and Travellers: The History, Development and Challenges of Gypsy and Traveller Tenants and Residents' Associations', Third Sector Research Centre Working Paper 84, University of Birmingham, UK.

## 2.9 CONNECTING MOTIVATIONS AND PRACTICE

To return to the practices summarised in Tables 1.1 and 2.1, and outlined in detail within the preceding sections, a tentative identification can be made of the particular models and discrete forms of practice that have satisfied the underlying aspirations of project initiators, activists and participants.

**Table 2.2: Connecting motivations to models of practice**

| Indicative motivation(s) | Indicative practice(s) to date |
|---|---|
| Building new property | Self-build, custom build, cohousing, land trusts, development trusts, co-ops |
| Providing local affordable housing | Land trusts, development trusts, co-ops, self-help |
| Property focused on rental provision | Co-ops, tenant management organisations, land trusts, self-help |
| Refashioning local services | Co-ops, tenant management organisations, community gateways |
| Renovating empty property | Self-help, co-ops, land trusts, development trusts |
| Low-impact/eco-dwellings | Self-build, ecovillages, land trusts, lowland crofting |
| Group living/utopias/political/ spiritual beliefs | Co-ops, tenant management organisations, 'intentional' groups |
| Creating shared/mutually supportive neighbourhoods | Cohousing, ecovillages, trusts, self-build |
| Moveable/mobile dwellings | Caravans, barges, canal-boats, tepees, vans, reconditioned buses |
| Stewardship by community anchor organisations | Land trusts, development trusts, co-ops, tenant management organisations, community gateways, self-help, ecovillages |

Table 2.2 provides indicative connections to understand how local people may have chosen to organise themselves to meet underlying needs and aspirations.

One of the inferences apparent in the table is that a plurality of factors can drive local activities. A single endeavour could also mix the construction of self-build units alongside other contractor-built properties, and all within a site that has a land trust as the long-term freeholder of accommodation let under a range of tenures.

**Table 2.3: Examples of complementary community-led motivations**

| **Project/ provider** | **Redditch Co-op Homes** | **New Ground Barnet** | **Tinkers Bubble** | **Canopy Leeds** | **Ashley Vale Bristol** |
|---|---|---|---|---|---|
| **Key model:** | Co-op | Cohousing | Eco-housing | Self-help | Self-build/ custom-build |
| **Motivations/ intent(s):** | | | | | |
| Building new property | √ | √ | √ | | √ |
| Affordable housing | √ | √ | √ | √ | √ |
| Controlled by occupants | √ | √ | √ | √ | √ |
| Management of existing stock | √ | | | √ | |
| Renovating empty property | | | | √ | √ |
| Low-impact eco-dwellings | | √ | √ | | √ |
| Communal operations | √ | √ | √ | √ | √ |
| Shared neighbour-hoods | √ | √ | √ | | √ |
| Community stewardship | √ | √ | √ | √ | |

One example of such a mix is the LILAC project in Leeds, with its mix of single-person and family-sized accommodation and other communal facilities, all constructed to a high ecological design and which describes itself as being a co-operative, a cohousing project and a community land trust.

No single 'model' is likely to be used in response to *all* of the motivations listed earlier, but a few examples can show how on-the-ground practices of different projects represent their responses to more than one underlying desire (Table 2.3).

What is considered in the next chapter is whether or not central and local government policies have understood the subtlety of the models discussed or the diversity of their intentions and their practices.

## Notes

1. For example, www.homebuilding.co.uk/market-report.
2. See www.hscht.co.uk/highland-self-build-loan-fund.html.
3. http://communityshares.org.uk/about-cooperative-and-community-benefit-societies.
4. See Nic Bliss (2017), *Mutuality and Accountability in the Housing Association Sector*, CCH, Liverpool, UK.
5. See www.uk.coop/students/about-student-co-op-homes.
6. National Association of Park Home Residents (www.naphr.co.uk). Independent Park Home Advisory Service (www.iphas.co.uk/index.php).
7. http://communityshares.org.uk.
8. http://westraydevelopmenttrust.co.uk.
9. Further details at www.letchworth.com/heritage-foundation.
10. https://locality.org.uk/.
11. www.communitylandtrusts.org.uk/what-is-a-clt/find-a-clt.
12. A recent publication from the Rural Housing Alliance mentions only CLTs as a community-led route to creating rural housing.
13. House of Commons Library (2019), 'Empty Housing, England', 29 May.
14. Details of funding allocations made by the Homes and Communities Agency under both rounds of the programme are available on the www.gov.uk website at: 'Empty Homes programme: guidance and allocations'.
15. Action on Empty Homes, 'Community Action on Empty Homes: Using Empty Homes to Regenerate Communities', www.actiononemptyhomes.org.
16. Wilson, W. (2017), *Council Tax: Empty Properties*, UK Parliament, House of Commons Briefing 00355, June, House of Commons Library, London.
17. See *Inside Housing*, 8 September 2017.
18. See interview with *Inside Housing* magazine, October 2017.
19. See www.lowimpact.org/lowimpact-topic/0-natural-building-intro/ for further information.
20. Blond & Briggs, London, 1973.

[21] https://grassrootsinnovations.org.

[22] See some histories recorded on the Diggers & Dreamers website: www.diggersanddreamers.org.uk.

[23] www.diggersanddreamers.org.uk.

[24] See also DETR (2003), 'Sustainable Communities: Building for the Future', ODPM, HMSO, London.

[25] CLG (2007), 'Ecotowns Prospectus', ODPM, HMSO, London.

[26] Demos (2015), 'Community Builders', https://demos.co.uk/project/community-builders-report/.

[27] See www.anglicancommunion.org/community/religious-communities.aspx.

[28] See www.bruderhof.com and www.northumbriacommunity.org.

[29] See for example, the annual *Retreats* publication issued by www.retreats.org.uk/index.php.

[30] https://www.diggersanddreamers.org.uk/resources/print.

[31] See www.diggersanddreamers.org.uk/communities/existing/redfield.

[32] See www.diggersanddreamers.org.uk/communities/existing/rainbow-housing-co-operative.

[33] See www.diggersanddreamers.org.uk/communities/existing/whiteway-colony.

[34] See http://differenttravellers.blogspot.co.uk.

[35] For example, www.travellerhomes.co.uk.

# THREE

# Enabling the creation of local homes: accountability or affordability?

## Understanding legislative interventions

Some broad comments were given in the opening to this publication on the UK's basic socio-political context in which its housing sector operates. These baldly stated:

- that established and mainstream participants are at the core of local frameworks that plan for how needs are assessed and met;
- that 'open market' practices currently dominate how innovations to services (and relationships) could be framed;
- that the extent of any local demand for different forms of housing solutions is assumed to be unproven, or may only be present on a very modest scale;
- that new community-based or self-provided solutions do not have the governance skills or practical expertise to take on the risks in driving new projects forward.

The previous chapter has, however, described in some substantial detail the kinds of innovations that local people have introduced to create the kinds of homes and neighbourhoods they want, developing new governance skills to drive projects forward that create sustainable homes for individual households, and created award-winning neighbourhoods and settlements in which new practices have been successfully integrated into daily and community life.

To what extent have such motivations and practices chimed with central and local policies? What will be considered is whether the motivations from local people to create local housing solutions currently feature in, or are supported by, 'mainstream' engagements. The consideration of these points will be principally undertaken though an examination of:

- the extent to which central government legislation and policy has promoted opportunities for people to influence how their homes and neighbourhoods are provided; and
- the shape and direction of policies at local government level.

Lastly, some consideration will also be given to the extent that support for local people to shape local homes and neighbourhoods currently involves an unresolved tension between any priority given to the *affordability* of local provision, and a focus on who is owed *accountability* for that local provision.

## Central government legislation and policy

As noted in the Introduction, amidst all the recent government policy imperatives to improve standards and delivery in the nation's housing sector and its housing supply,[1] there has been a sustained focus on augmenting the supply from established providers (such as private sector housebuilders or housing associations), rather than on generating changes in market conditions to encourage other activity to take place. That has not, however, meant central policy has been silent on ways to stimulate some degree of new provision from other sources, such as contributions from local communities and new local groups.

Recent central policy initiatives include statutory powers from the Localism Act 2011 to increase local opportunities for communities to undertake 'small-scale, site-specific, community-led developments'[2] – the kind that can dovetail with the government's 'National Housing Strategy' and promote distinctive development designs 'that reflect local character and identity'.[3] A national 'Self Build Action Plan'[4] was produced hoping to double the output of the self-build sector, and the 2012 publication of sweeping revisions to the national and statutory planning framework – the 'National Planning Policy Framework'[5] – carried a requirement on planning authorities to assess the local demand from people who wish to 'build for themselves'.

Subsequent policy developments have sought to extend help to make sites available for community builders; to provide 'seed' finance to cover initial scheme development costs; and to reduce potential obstacles from bureaucratic 'red tape' previously applied to small building projects (including an exemption for self-build projects from local Community Infrastructure levies or to meet Section 106 planning obligations of the Town and Country Planning Act 1990).

These were followed by the Self-Build and Custom Housebuilding Act 2015 and the Self-build and Custom Housebuilding (Register) Regulations 2016, which have subsequently placed requirements on English authorities to introduce registers that will 'have regard to individuals/ bodies who have expressed an interest in acquiring land for self-build purposes'. Authorities should now consider how the identification of 'serviced plots' could be provided in response to such aspirations (although the regulations do not state that the local authorities must provide such plots, merely to identify the possibility of where they might be found).

Further policies containing fiscal and investment incentives have included:[6]

- a custom-build investment fund;
- a loan fund to help provide custom-build serviced plots;
- support towards self-build mortgages;
- elements of the programmes using public grant towards general housebuilding initiatives (now managed through Homes England).

It should however be acknowledged here that some of this investment – like the Home Building Fund – has assisted traditional housebuilders in their options for what to supply to private households, rather than resourcing aspirant households direct.

The latest initiative, the Community Housing Fund, commenced late in 2016 and significantly extended from 2018, has provided revenue and capital funding for specific community-led projects and for the development of national support frameworks that can embed such activities into local supply networks. The allocated sum between summer 2018 and March 2020 was in the region of £160m, more than has previously been earmarked for community-led initiatives, although still a fraction of the billions routed towards the funding of more conventional housebuilding activities through the mainstream housing development firms.

2018 also saw the government issue a 'Social Housing Green Paper'[7] promoting a variety of potential reforms, including an increase in the empowerment of social housing tenants to obtain 'good quality and safe social homes with the right services from landlords', although the parliamentary focus on European matters took attention and momentum away from the substance of those original consultations. Other recent government initiatives could, in theory, support the implementation of community housing projects – such as via procurement opportunities framed by the use of the Social Value Act 2012,[8] or through the kinds of collaborative frameworks 'to build thriving communities' promoted within the recent Civil Society Strategy 2018[9] – however central policy has yet to lay down any requirements that would enable local communities to feel confident in applying such measures to decide how housing and neighbourhood development opportunities are implemented.

What has been central to recent imperatives could therefore be summarised as:

- instructing local authorities to be more precise about the local demand and desires being expressed by local people to create their own homes;
- implementing new political opportunities (legislative 'rights') through which local communities may influence local outcomes and the use of local resources;
- providing funds to help the practical implementation of local initiatives (such as via the previous Empty Homes Programme, and the Community Housing Fund).

The rationale evident in such recent policy is, however, more to support ways whereby local communities and their households might increase the overall volume of new housing being built, rather than help local community projects challenge the position of key interests within the UK's housing and development sector. In particular the range and focus of government policy initiatives has not facilitated any real challenge to the possible monopolisation of planning and land assets in the UK,[10] which has become something of a litmus test for assessing how local grassroots initiatives may or may not progress.

A series of academic and research reports[11] have highlighted severe obstacles in acquiring sites and obtaining development finance that prevent take-up of local and household-led housing initiatives. Ted Stevens has noted the routine work by the National Custom & Self Build Association to highlight what self-builders experience as their biggest barriers – more than 50% in a recent survey had experienced 'finding land' as their greatest challenge.[12]

**Figure 3.1: Challenges to self-builders**

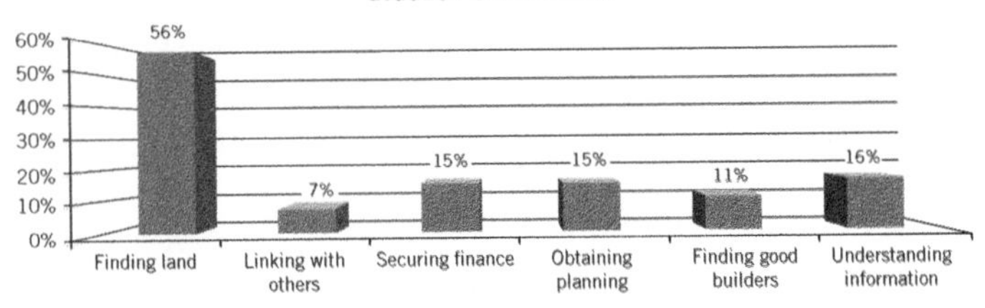


Source: T. Stevens / © Benson & Hamiduddin 2017

Yet the impact of obligations on planning authorities to identify local demand, for example as set out in the National Planning Policy Framework and in the requirements of the national Self-Build Regulations, is yet to prove substantial. Many current local authorities are only producing a record of those households keen to build or provide their own dwellings – there is no obligation on them to make land available on which aspirant households could build for themselves. The regulations state that plots should be identified, but then do not provide the means for willing households to acquire them.

Even where some planning authorities are making reference in revisions to Local Plan frameworks for the future provision of 'self-build/custom build plots', what is emerging is a linkage of future obligations to the possible phases of larger residential schemes, which may never materialise. One recent formulation states that 'the Council will seek the provision of land for custom and self-build housing on housing allocations capable of providing 250 or more dwellings, as part of an appropriate mix of dwellings'.[13] Such policies will do little to assist aspirant self-builders in finding a site in smaller developments (smaller than 250 units) or to make use of a plot within bigger schemes in any convenient timescale: if the provision of plots is tied into a later phase of a large development that could be years into the future.

The Regulatory obligation is certainly not being interpreted as a practical tool to help the households recorded in the Self Build Register acquire building land. Rather it is being linked to abstract or potential connections back to the main housing providers whose practices on land-hoarding and slowness in undertaking site completions are increasingly seen to be self-serving rather than supportive of wider public policy.[14] It remains a concern that the national policy framework can permit such local interpretations which, on paper, seem to acknowledge a positive response to national imperatives, but will in practice constrain future opportunities for how grassroots aspirations might be progressed.

## Policies and practice at local government level

In reviewing how local authorities might have recognised a role in supporting local communities to create their own homes, academic consideration[15] has included a look at what might be recognisable within formal strategic frameworks that have been shaping local property and residential development. A series of desktop research studies conducted through 2014–17 examined a variety of public sector housing and planning strategies pertaining to development in the largest urban settings in England. This searched for any explicit or evident reference to future housing delivery taking place through 'self-provided' or 'collective' routes of action. The examination sought to identify reference to 'mutual' or 'collective' or 'self-provided' activities to develop local homes and/or neighbourhoods – specifically it considered if there was any support or mention for any of the following activities: self-build; custom-build; self-help; co-operative; community land trust; cohousing; and/or community-led housing.

A summary of the results of these studies has been included within the Appendix at the end of this publication. What can be noted here is the general lack of formal support found at the time for most of the activities highlighted, regardless of the overarching imperatives that have come from central government. Admittedly, some of the documents were from frameworks in place for several years and which had been drafted before some of the government's emerging community-led policies came into effect, although the authorities would still have had time to produce supplementary 'sub'-strategies or other dedicated policies that could relate to community-led or self-build housing.

Even where some community-led projects were known to be taking place, they seemed to exemplify idiosyncratic 'one-off' projects, driven by the persistent efforts of individual households or community groups, rather than being outcomes of a definite local policy intent, or of a kind that could be easily replicated.

What emerged from that study was the sense of there being a limited appreciation within the public sector for how it might use formal frameworks to support opportunities for innovative work undertaken by self-organised or community-led housing initiatives. Some frameworks acknowledged a hazy awareness of 'self-build' schemes, or of new 'community land trusts', but overall there was little depth of understanding of, or support for, the wider range of community intents and capabilities that have been explored here.

This can be contrasted with more practical advice that is seeking to guide public sector support towards practical ways that local community-led initiatives could be encouraged and resourced. Both the 'technical toolkit' produced by the Housing Association Charitable Trust[16] and the publication 'Community-Led Housing: A Key Role for Local Authorities' produced by the 'Commission on Community-Led Housing',[17] sponsored by the Co-operative Councils Innovation Network, detail how to support local and grassroots projects in sustainable and appropriate ways. Both underline the benefit of statutory and professional bodies adopting meaningful local policies to support the community-led sector, such as the Supplementary Planning Document to support community-led projects from East Cambridgeshire District Council,[18] and the advice to professional bodies and practitioners within the 'Planners Guide to Community-Led Housing',[19] designed 'to provide planners who work in England with the knowledge and confidence to devise policies and adopt development management practice that supports CLH delivery'.

It is, however, still clear from the relative scarcity of in-depth policies that more precise examples of local authority strategies and policies are waiting to be formed, especially if broad community motivations are to be acknowledged, respected and resourced. And over and above any creation and adoption of policies that may impact upon individual aspects of local services and community aspirations – such as about land supply, or the nature of affordable housing, or acquiring empty properties, or the use of design codes – a headline position should be taken at the highest level of local authorities that will value community initiatives *in their own right.* A wholehearted response to communities wanting to be empowered to help themselves would not squeeze them into marginal positions of what might be provided by others, but would engage with local residents and tenants direct.

A set of suggested issues that could form the substance and format of such future public sector policies and strategies is therefore proposed in Table 3.1, along with ideas on the kind of impact(s) they could have.

## The 'affordability' and 'accountability' of community-led provision

The previous discussion on central and local policy developments has repeatedly touched on the *affordability* of housing provision.

At a broad interpretation, the concept of housing being 'affordable' might be viewed as somewhat subjective, in particular depending on the income and resources the prospective occupant(s) have available to meet their housing expenses – rent or mortgage or something else. A self-build project constructed for ownership by a low-income household might still be considered *un*affordable if a household's income is only at the level of statutory benefits, and no other funds are available to meet any loan repayment costs in the future.

**Table 3.1: Forms of local authority policies to support collaborative housing projects**

*(3.1a) Baseline positions by local authorities to support grassroots initiatives from local people to steer the creation of local homes and neighbourhoods*

| LA responsibility | Indicative or potential policy | The 'community-led' purpose within future policy |
|---|---|---|
| **Statements for corporate and community engagements** | Community Strategy, or similar statement | To champion 'community-led' place-making and initiatives |
| | Joining partnerships for joint activities with external bodies | 'Cross-departmental' teams helping external projects |
| | Implementing the powers of the Social Value Act | Procurement opportunities for community-led enabling hubs |
| | To set up projects for people to create their own homes | Employment or secondment of 'enablers' to assist projects |
| **Implementation of 'community rights'** | A process to designate 'assets of community value' | To note land and properties to safeguard for community use |
| | The creation and content of neighbourhood plans | To encourage plans to include community-led projects |
| | The potential focus of community 'bids' and 'claims' | To note services community projects might wish to 'claim' |

*(3.1b) Policies to help focus strategic and local planning initiatives*

| LA responsibility | Indicative or potential policy | The 'community-led' purpose within future policy |
|---|---|---|
| **Strategic spatial plans and development control/ development management** | Local plans/core strategies | To detail sites where residents could lead on creating homes |
| | Strategic land availability assessments | Assessment of sites for resident-led/community use |
| | Supplementary planning documents | To expand on the detail for local community-led outcomes |
| | Specific design codes | Local standards for housing/neighbourhood change |
| | Specific development orders | Procedures to support housing/neighbourhood change |
| | Requirements for, and delivery of, 'affordable housing' | To dovetail such delivery with community-led proposals |
| | Exception site policy | To encourage community-led proposals to lead 'exceptions' |
| | Policies to meet demand on local 'Self Build Registers' | How registered households may progress ideas to build |

*(3.1c) Policies to shape the way local authorities apply their powers and assets in support of local and collaborative housing projects*

| LA responsibility | Indicative or potential policy | The 'community-led' purpose within future policy |
|---|---|---|
| **Corporate asset management** | The sale or use of local authority land and buildings | To frame the release of assets to community initiatives |
| | The role of the authority to implement site infrastructures | To release 'serviced' land to individuals and groups |
| | To lead by example | To designate sites as 'key demonstrators' |
| | To facilitate take-up of public and private sites | To set up a 'small sites portal' for pre-application searches |
| | Liaison with other public sector bodies | To establish wider support for new community projects |
| | Acquisition of other land/assets to meet core strategies | To be proactive in finding land for citizen-led projects |
| **Corporate finances** | Use of authority finances to implement corporate strategies | To provide community projects with 'start-up' finance |
| | Use of corporate skills to aid wider financial modelling | To assist community projects capital and revenue plans |
| | Ability to give guarantees for mortgage/loan applications | To assist projects to secure finance and reduce risk |
| | Discretionary grant finance to help 'enable' external bodies | To allocate developmental grant to enable CLH initiatives |
| **Legal services** | To oversee legal frameworks of local authority engagements | To set realistic expectations from local community bodies |
| | To review governance details of partnering bodies | To note formulas for agreed community-led incorporations |

*(3.1d) Policies that focus the strategic approach of housing and regeneration work to incorporate projects from local individuals and community groups*

| LA responsibility | Indicative or potential policy | The 'community-led' purpose within future policy |
|---|---|---|
| **Housing and regeneration strategies** | Strategic priorities within core housing strategy documents | To detail what community-led outcomes can be facilitated |
| | Strategic housing market and housing needs assessments | Assess the local demand for community-led outcomes |
| | | Undertake sufficient review of services for travellers |
| | | Assess other demands: living in groups, on barges, etc |
| | Content of 'homelessness' strategies | Projects for households with no homes/insecure tenancies |
| | Operation of the statutory 'Housing Register' | Support for community-led ways to meet 'affordability' |
| | New-build/rehabilitation projects in 'regeneration' areas | Use of capital for community-led redevelopments |
| **Procurement of new housing supply** | Procurement panels to commission new-build supplies | How to award contracts to 'community-led' bodies |
| | Frameworks for the supply of affordable housing | Awarding grants to schemes that deliver core targets |
| | Other help and advice to implement approved projects | To establish a 'small builders portal' of available providers |
| | To resource opportunities to develop skills and expertise | To resource communities to improve project management |

*(3.1d) Cont.*

| **Social sector housing management** | Management of council housing | Promotion of opportunities for tenant management bodies and other community-based initiatives |
|---|---|---|
| | Alignment of housing stock in regeneration proposals | |
| | Codes of practice for stock from stakeholders | |
| | Local implementation of any 'Right to Buy' legislation | To negate reduction of any collective freehold ownerships |

*(3.1e) Policies applied to the wider revitalisation and management of neighbourhoods and other localities*

| **LA responsibility** | **Indicative or potential policy** | **The 'community-led' purpose within future policy** |
|---|---|---|
| **Environmental health** | Strategy for action on empty homes | Property acquisitions coming to community-led bodies |
| | Use of Empty Dwelling Management Orders | Leasing of properties for long-term use by community bodies |
| | Management of private rental sector provision | Communities managing stock to deliver to policy standards |
| **Landscape design and management** | The development of high-quality 'place-making' | To enlist local communities to contribute to design initiatives |
| | Management of public realm and open space | A role for community bodies to undertake area 'stewardship' |

In the UK, the more customary use of the concept of 'affordable housing' is the description given to the dwellings and tenures of local authority housing services, or from housing associations and from other social housing providers (now also including contributions from housing co-operatives and community land trusts). Due to the changing nature of the UK's housing market, the proportion of affordable housing is smaller now than both the owner-occupation and private rental parts of the wider housing market and is heavily dominated by services from local authorities and housing associations. The dominance of those agents is reinforced by wider planning, construction and investment relationships that have emerged over time, not least in the focus of the central government investment and regulatory regimes used to direct state resources towards this segment of the wider sector, which is especially focused on providing help to households 'unable to meet their own needs through the open market'.[20]

The linkage between state resources and how these could be directed towards underlying elements of housing need has become a central factor in how many community-led and self-build initiatives are now viewed or interpreted by those large-scale providers and statutory bodies. This is regardless of whether or not there might be a perceived overlap (or duplication) of interests for the allocation of resources to help provide forms of local affordable housing, and whether or not the community-led initiatives might offer to fill a gap in local services not being covered by others.

The focus of much statutory investment in local community housing schemes has reinforced this factor. Whilst the Prospectus of the Community Housing Fund invited all kinds of community-led proposals for new or existing forms of housing provision, it still linked this to a requirement that future use of public funds be directed towards some provision of *affordable* housing – schemes would be unlikely to get any significant support if based solely on other tenures.

Whichever interpretation of affordability is used – whether to meet the social housing sector's definition of 'need', or simply to suit local incomes – local people and their communities invariably focus upon creating homes in which people can properly afford to live. Some proponents of community-led schemes are indeed ideologically committed to a position that community-led properties should be focused solely on local supplies that are accepted as affordable housing by the established mechanisms, particularly when providing property for rent. This has been a basic position of the housing co-operative movement for many years and is being reinforced by the principles of new community land trusts, that housing provision should be provided as 'affordable in perpetuity', in order to benefit successive occupants.

There has, however, been an unfortunate consequence of this association or presentation of UK community-led projects as affordable housing that a standard appreciation of the community-led sector has emerged that now sees it as only a sub-set of the affordable housing sector, which is itself is recognised as just a sub-set of the wider housing sector. It is clearly through this kind of lens that many formal bodies inside the housing and development sectors now respond to ideas coming from grassroots interests: proposals are assessed in terms of how understandable or justifiable they seem, viewed by mindsets that will interpret all aspects of community-led projects as needing to demonstrate new provision of an identifiably affordable housing nature. The UK Cohousing Network was informed by the executive management level of the (then) Homes and Community Agency that the perception of cohousing was of a tenure that the agency would not support in any great degree, as its proponents do not seem to fit prevailing stereotypes of households 'in housing need'. More than one community-led proposal has encountered the response that people advocating solutions for themselves are not in any outstanding need (and may even be stereotyped as 'middle class') and thereby are not eligible for support being legitimately supplied from any public resource (in the form of finance or as policy), so support from those statutory sources will be directed elsewhere.

Even where community-led initiatives are adamant that they wish to promote and procure opportunities for local affordable housing of the kind that local authorities or housing associations would otherwise provide, initiatives can be viewed by mainstream bodies as seeming too selective in their local approach and perhaps not sufficiently 'representative of the wider community'. Representatives of the local authority sector are known to claim that the UK's background to the local electoral mandate is the only legitimate context on which to assess how local provisions should be framed. Aspirations and ideas put forward by identified households, such as for self-build projects, are translated into representing submissions for personal advancement which the statutory sector need not support, not least because these do not fit with the *affordable* stereotypes.

Notwithstanding the community benefit that will always come from creating local homes that are genuinely affordable for their occupants, there are clear negatives in equating community-led projects with just a creation of 'affordable' housing (not least the risk of being swallowed by avaricious external bodies that can and do claim that their own level of operations will be more cost-effective). If the greater percentage of UK households still aspire to some form of property ownership, as MORI opinion polls regularly find, then a promotion of local projects only to provide properties for rent will not engage many of them. The 'not-for-profit' or 'community benefit' ethos of a community-led scheme should be cautious of presenting itself, or of being presented by others, as being synonymous with solely rental or other mainstream affordable tenures. They will also benefit from being seen to countenance and promote forms of collective and leasehold ownership, and schemes that provide for a mix of tenures.

Members of local initiatives can have a variety of aspirations, and it is legitimate for schemes to present their plans for a variety of tenures and ownership outcomes, if that is what their members wish. Too often it is as if community-led projects think they must present themselves as able to provide affordable housing in order to gain credibility with potential mainstream partners, but then are assessed as less capable of doing a good job than those other bodies. There is a wider principle here of giving support to a scheme promoted by a local community or community group, because it will have intrinsic worth in having been put forward by the local households concerned.

The contemporary focus on affordable housing provision from the community-led sector being of a comparable nature with what local authorities and housing associations provide is increasingly obscuring a wider appreciation of how to value and respond to the desires of local people to shape their own homes and neighbourhoods, *regardless of tenure*. A basic level of support for projects to be accountable to their initiating households, regardless of ownership or tenure issues, is being subsumed under other values that have arisen from concerns about the affordability of future properties – concerns which have then been commandeered by mainstream technical frameworks for what should happen to 'affordable' properties, to the potential detriment of community-led controls. A very definite re-evaluation of 'accountability' is required here.

The prevalent impression within the housing sector is also that most community-led practices are small in nature and in their settings. This becomes a constriction to having a proper appreciation of what grassroots projects might achieve, if they only had commensurate opportunities of the kind available to mainstream providers. The next chapter looks at some practical examples of community-led success elsewhere in Europe that can remove any doubt about the scale at which local people have been shaping some very significant social environments.

## Notes

[1] Wilson, W. (2019), 'Briefing Paper SN06416: Stimulating Housing Supply: Government Initiatives', House of Commons Library.

[2] Department of Communities & Local Government (2011), 'Localism Act 2011', Part 5 Community Empowerment, http://www.legislation.gov.uk/ukpga/2011/20/contents/enacted.

[3] Department of Communities & Local Government (2011), 'Laying the foundations: a housing strategy for England' (p. 57).

[4] National Self Build Association (2011), 'Report of the Self Build Government-Industry Working Group', see https://issuu.com/simplymarcomms/docs/nasba_action_plan.

[5] Department for Communities and Local Government (2012), 'National Planning Policy Framework', HMSO, London.

[6] Wilson, W. (2017), 'Briefing Paper SN06784: Self Build and Custom Build Housing (England)', House of Commons Library, London; Wilson W. (2019), 'Briefing Paper SN06416: Stimulating Housing Supply: Government Initiatives', House of Commons Library, London.

[7] Ministry of Housing, Communities and Local Government (2018), 'A New Deal for Social Housing', HMSO, London.

[8] HM Government (2012), 'Public Services (Social Value) Act', HMSO, London.

[9] Office for Civil Society (2018), 'Civil Society Strategy: Building a Future that Works for Everyone', HMSO, London.

[10] As raised in the 2017 Housing White Paper.

[11] See Wallace, A., Ford, J. and Quilgars, D. (2013), 'Build-it-yourself: Understanding the Changing Landscape of the UK Self-build Market', Centre for Housing Policy, University of York; Heywood, A. (ed) (2016), *Local Housing, Community Living: Prospects for Scaling Up and Scaling Out Community-led Housing*, Smith Institute, London.

[12] Stevens, T. (2017) 'Turning the Theory into Reality', in Benson, M. & Hamiduddin, I. (eds), *Self-Build Homes: Social Discourse, Experiences and Directions*, UCL Press, London.

[13] Harborough District Council (2019), 'Harborough Local Plan 2011–2031' (adopted April 2019).

[14] Archer, T. & Cole, I. (2016), 'Profits before Volume? Major Housebuilders and the Crisis of Housing Supply', Centre for Regional Economic and Social Research, Sheffield Hallam University, UK.

[15] Field, M. (2017), 'Models of Self-build and Collaborative Housing in the UK', in Benson, M. & Hamiduddin, I. (eds), *Self-Build Homes: Social Discourse, Experiences and Directions*, UCL Press, London.

[16] Housing Association Charitable Trust (2018), 'Community-Led Housing Technical Toolkit', HACT, London.

[17] CCIN (2018), 'Community-Led Housing: A Key Role for Local Authorities', CCIN Network, UK.

[18] East Cambridgeshire District Council (2016), 'Community-Led Development Supplementary Planning Document (SPD)', https://www.eastcambs.gov.uk/sites/default/files/Com%20Led%20Dev%20SPD%20as%20adopted%2025%20Feb%202016.pdf.

[19] Lavis, J. (2019) 'Planning for Community Led Housing', https://communityfirstyorkshire.org.uk/plannersguide/.

[20] An underlying principle for public sector intervention within UK planning and housing provisions.

# FOUR

# Learning from Europe: building at larger scales

## Examples of large-scale developments

There is tendency to bracket contemporary UK community-led aspirations with a belief that these are going to be quite localised, principally modest projects for a few new 'affordable' units, in a very localised setting (provided by a body like a community land trust), which will be small in scale and rural in nature. The following examples should offer grounds to reassess how community-led initiatives have been coping with significantly larger scales of development.

A single example to take from the UK is the current Graven Hill housing development in Bicester, where the local authority has responded to local people's obvious appetites to be involved in planning and building new homes. The authority has acquired land on which it is currently planning for 1,900 properties to be delivered through a variety of forms of local and self-directed activity. The first phases of individual plots have been developed through self-build and custom-build developments and some structured development by housing associations and private developers could be taken up in the future. Initial plot prices are reflective of a buoyant local market, so it is not yet an obvious route for lower-income households to access the project. However, it is hoped that opportunities will materialise in due course.

Both in spirit and in substance, the Bicester approach has drawn from experience developed in town and city development on the European mainland, particularly from the suburban development established by the local authority and its partners at Almere in the Netherlands. Examples of high-quality community-led housing developments are now identified across Europe, and compilations like the *European Co-operative Housing*[1] in 2015 pull together exciting experiences of collaborative developments across a range of countries (including some schemes from the UK).

What is illuminating within many continental projects are the core principles that challenge the manner in which large-scale speculative UK housing and neighbourhood developments are usually undertaken, providing accommodation for households that have yet to be identified nor engaged in any subsequent practice of neighbourhood creation. Continental examples offer some very real alternatives to this.

Four different European examples will be cited here:

- the *Egebjerggaard* urban extension in the Greater Copenhagen area in Denmark;
- the '*Baugruppe*' neighbourhoods at Vauban in Freiburg, Germany;
- the self-build and custom-build expansion area of *Almere* in Holland; and
- the *Meer als Wohnen* co-operative neighbourhood, in Zurich in Switzerland.

### *Egebjerggaard urban extension, Denmark*

As part of its planning for significant new areas of urban growth, the municipality of Ballerup to the north of the Copenhagen municipal region on Sjælland (Denmark) adopted a town planning policy to provide significant opportunities for local people to shape how some new settlements could develop. The policy particularly sought innovative ideas that would be a challenge to conventional property developments that were increasingly seen as being boring and featureless.

Operating in partnership with Denmark's main state research body on architecture and the built environment, Ballerup engaged a select group of architectural firms to collaborate on devising key planning and design principles for how the new settlement of Egebjegrgaard could be created as a suburb of approximately 900 homes through the late 1980s and 1990s. This collaboration led to proposals for core planning and design concepts of how the suburb might be constructed; how its phased development could be managed; and to strengthen the social relationships that could be fostered within its borders. The subsequent major phases of development were each competitively tendered to the property and design market, with the winning tender responsible for producing an initial masterplan for each phase. Subdivisions of each master-planned area were themselves designed and developed by a wide range of approaches. Some parts were acquired by entrepreneurial housebuilders; some by cohousing groups; and some by other kinds of co-operative groupings. Most of such groupings were comprised of future residents, building or designing for themselves. Individual competitive designs were also commissioned for the community facilities to complement the main residential provisions of each phase. Each area's residents decided on the nature of the facilities that would support a shared 'urban life' in their local neighbourhood.

Revealingly, the winning tender of the first 'demonstrator' phase was the architectural practice steered by Jan Gudmund-Høyer, one of the founding members of the Danish cohousing movement, which had also been one of the original key contributors of ideas for local suburban planning policies and their values and standards.

The resulting urban environment seamlessly weaves together dwellings and open spaces, and areas of privacy and communality, even in those parts which were developed and sold as more traditional 'speculative' housing, adjacent to other properties pre-commissioned by prospective residents. Property tenures were mixed throughout the settlement, and throughout its constituent phases. This was assisted both by the public funding available at the time, and by the main principles of the planning policy, which shaped a very inclusive set of collaborative opportunities between all kinds of households and socioeconomic groups.

### *Vauban 'Baugruppe' neighbourhoods, Germany*

Vauban lies 3 kilometres to the south of Freiburg city centre, a small university city (population 225,000) in south-west Germany, near Switzerland and France. It is a suburban area readily accessible by tram from Freiburg city centre (a 15-minute journey). The means of creating an urban extension principally of community-led projects was conceived in the mid-1990s. New-build development began on site in 1998 and was completed by 2002.

The 'Baugruppe' represent a mixed-use suburban setting, built on brownfield land at the edge of the existing conurbation, housing 5,000 people in approximately 1,800 units (plus 600 units of student accommodation) and providing some 600 jobs. Facilities include a primary school, kindergartens, shops, supermarkets, community centre, market square, children's play spaces, sports field, various local services, plus some cafes and restaurants. The neighbourhood is served by a tram running along the main avenue, which became operational in 2006.

The site was a former military base, which the city council bought in order to pursue a planned policy-led development of the new suburban area. The overall development was driven by the city council, led by the chief planner Wulf Daseking and by a citizen's group called Forum Vauban, which was made up by early residents living in or close to the site prior to the project's commencement.

A masterplan was created by the municipality with a prescriptive design code and regulating plan, which set out a precise set of rules for phases of development. Individual building developers submitted their own plans for sub-neighbourhood development in order that these would conform to the design code and meet associated housing policies. The authority then acted as 'head' site developer and took responsibility for installing the extensive public infrastructure across the area. It then sold serviced plots to the 'commissioning bodies' (acting as their own local developers). Blocks of land were targeted towards 'Baugruppe' – independent co-operatives of would-be resident households. These could direct their appointed agents to design their blocks and local neighbourhoods, planning for properties that would accord with the main design code and with household requirements.

Whilst the overarching housing policy wanted a mix of tenures, with at least 30% of housing being for rent, a number of 'Baugruppe' subsequently prioritised the delivery of properties for affordable home ownership and obtained agreements to lower the final percentages of dwellings for rent.

### *Almere 'self-build' urban development, Holland*

The large-scale development of this 100-hectare space lies to the south west of the planned city of Almere, on land reclaimed from the sea in the 1950s. It is part of the long-term planned expansion of the city, with areas master-planned by the local authority into several districts, each of which has around 720 self-build plots. Eventually some 3,000 self-built homes are planned.

The self-build districts have been 'themed': there are areas for living/working homes; there are terraced areas; and there is a zone for high-quality 'sustainable' homes. An area is set aside for canal-side homes, and another for houses with substantial gardens. There is also a zone specifically aimed at housing developers, commissioned to build apartments or groups of terraced homes.

The infrastructure of roads and utilities is installed by the council, and then serviced plots are purchased by households directly from the local authority. These vary in size from 86 square metres to more than 1,200 square metres, with an A4-sized 'plot-passport' that acts as the permit to build. This specifies the main conditions applicable to any construction, though it can also govern its height, any required gaps between neighbouring homes, and any building line for the front and back alignment of the future properties. Beyond this, the plot owners have free choice in the design, materials and appearance of their homes.

Many have opted for a modern, contemporary feel and several futuristic-looking homes have been constructed. Others have chosen to use more standardised kit homes, and there have been a variety of approaches, where some work is done by a contractor or developer and some is completed by self-builders. Currently little new-build activity has been conducted by the main volume housebuilders, yet overall self-build activity has still proceeded quickly.

The plot developments have furthermore involved a wide diversity of people of varied ages – from teenagers through to people in their seventies, and from people across a range of cultures. Not all the homes are built as solo projects – 25 households formed a group and hired an architect and contractor to build a block of flats together. This enabled them to get their flats at highly competitive rates in comparison to the open market.

### *Mehr als Wohnen co-operatives, Switzerland*

Switzerland's political aim of making housing 'inclusive and affordable' has created a supportive environment in which the capital city of Zurich offers all public land to local co-operatives before it is offered to other property development bodies. Co-operatives are also able to expand through the city council buying underused sites and offering them to co-ops on long-term leases. Financial support towards housing costs in Zurich includes 'rent help' and a 'solidarity fund' (deliberate terminology used in preference to a phrase like 'housing benefit') that has enabled residential participation by a wide range of households. In 2011 the city held a referendum on the direction of its future housing provision, and this resulted in a vote and commitment to increase the proportion of not-for-profit housing to 33% of *all* housing by 2050 (not just newly built homes).

*Mehr als Wohnen* (translated as 'More than Housing') is a co-operative community of approximately 1,400 residents in high-quality mixed-tenure accommodation. This formed a new quarter of Zurich on its completion in 2015, housed within 15 minutes of the city centre. A recent book on the scheme[2] noted that the inspiration for the project was to celebrate how co-operative and not-for-profit housing has 'decisively shaped the character of the city' for over 100 years.

In 2017, *Mehr als Wohnen* won the top prize at the internationally assessed World Habitat Awards, run by World Habitat (formerly the UK's Building and Social Housing Foundation). The prize recognised the focus on community engagement by its 13 blocks of residential buildings, forming two enclosed public squares and streets. Ground floor space is dedicated to a variety of commercial and communal uses. These include two restaurants, a communal kitchen and bar, a classroom, and studios for art, music and other leisure activities.

## Lessons from large-scale community-centred developments

A number of points can be gleaned from these examples about what is identifiable at the core of their respective success:

### *Political will*

The schemes cited all had clear and high-level political support for the approach they had taken.

Each of the schemes demonstrated a prior belief in the positive impact of supporting local people to build their own homes within large-scale development programmes.

There was a demonstrable accountability of the outcomes, with the use of the land and available resources resting with local people and their communities.

### *Enabling practices*

Each of the schemes ensured that sufficient land would be available for the headline purposes they put forward.

The public authority took responsibility for implementing the infrastructures that would service each development area and frame the character of its building plots.

Teams of professionals were organised by the municipal authorities to facilitate how local people would make use of the resources necessary to implement development.

There was a readiness to adopt planning and development guidelines that support diversity in property and neighbourhood designs, and a flexibility to work on-site.

The implementation of local decisions and codes for property and neighbourhood designs were ultimately in the hands of future residents.

### *Inclusive outcomes*

Each of the schemes was accessed and used by a broad range of households and community interests.

The development in all the areas demonstrated a breadth of linkages and interaction between residential use and other local and community facilities.

The costs of plot development in each scheme have proved more affordable and deliverable than those of speculative private sector construction.

## Notes

[1] Wohnbund e. V (2015), *European Co-operative Housing*, jovis-Bücher, Berlin.

[2] Hugentobler, M., Hofer, A. & Simmendinger, P. (2015), *More than Housing: Co-operative Planning*, Birkhauser, Basel, Switzerland.

## Further reading

Atwell, K. (2002), *Liv i Egebjerggård, Resultater 022* (ed.), Danmarks Statens Byggeforskningsinstitut, Copenhagen, Denmark.

Ballerup Kommune (1989), 'Bykvarterkatalog: Østerhøj og Egebjerggård', Ballerup, Denmark.

Bektas, E., Brouwer, J., Di Giulio, R., Pasqualis, M.B. & Quentin, C. (2014), 'A Suitable Design Methodology for Collective Self-organised Housing Projects to Build Sustainable Districts', *Sustainable Building 14, World Congress*, 28–30 October, Barcelona, Spain.

Czischke, D. (2017), 'Collaborative Housing and Housing Providers: Towards an Analytical Framework of Multi-stakeholder Collaboration in Housing Co-production', *International Journal of Housing Policy*, published online 20 June, https://doi.org/10.1080/19491247.2017.1331593.

Davies, D., Snelling, C., Turner, E. & Marquardt, S. (2017), 'Lessons from Germany', Institute for Public Policy Research, London.

Hamiduddin, I. & Daseking, W. (2014), *Community-based Planning in Freiburg, Germany: The Case of Vauban*, Policy Press, Bristol, UK.

Housing Europe (2017), 'The State of Housing in the EU 2017', European Federation of Public, Cooperative and Social Housing, Brussels.

Lang, R. & Harald Stoeger, H. (2017), 'The Role of the Local Institutional Context in Understanding Collaborative Housing Models: Empirical evidence from Austria', *International Journal of Housing Policy*, published online 7 February, https://doi.org/10.1080/19491247.2016.1265265.

O'Leary, J. (2014), 'Self-build Britain: Is the UK Lagging behind Other Countries?', *The Independent*, 24 March.

Pittini, A., Laurent, G., Julien, D. & Kiss, I. (2015), 'The State of Housing in the EU', European Federation for Public, Cooperative and Social Housing, Brussels, Belguim.

Sakula, A. (2013) 'Around the World: Volume House-building Is not Normal', University of Sheffield, www.ashsak.com/projects/collective-custom-build.

Wolf, M. (2017), 'Diversifying the Housing Market through Custom- and Self-build', *Town & Country Planning*, June 86(6): 259–65.

# FIVE

# Evaluating impact in a 'broken market'

## Setting benchmarks

Given the preceding elaborations of what is sought in local housing and neighbourhood projects, what could make their success more or less likely? In the context of current government concerns that the UK housing market represents a faulty and 'broken' set of arrangements, what impact could more local actions and initiatives achieve? And against what benchmarks might future success be assessed?

If the core critique of the 'broken' market is that many properties and tenancies are still too expensive and too few in number, then community-led and self-build projects will clearly have a positive impact whenever they provide local dwellings that are not forthcoming from other sources, and at affordable prices. If a fuller consideration of the wider market suggests there is a persistent or even increasing monopolisation of UK land and housing resources by larger influential and self-serving interests, and that these exclude local households from participating in key decisions over local provisions, then *success* for the promotion of community-led and household-driven projects will be judged in terms of whether or not local provisions are to become more accountable to those who will reside in any homes created.

Reference has been made in earlier chapters to the kinds of innovations in central and local government policies that could exert greater leverage to help collaborative and self-build housing schemes exercise more local influence. Future changes in the actions and responses of statutory bodies will necessarily feature within conceptions of indicators that will evidence a positive impact from the community-led housing sector's activities.

A first list of what such indicators could be are given in Table 5.1, contrasted against a note of underlying challenges that confront the community-led sector's current influence.

Key to how the potential impacts listed in Table 5.1 could materialise over time will be the number and quality of collaborative and self-build projects that emerge from local initiatives. Yet if the level of community-led housing initiatives has historically been modest (at least in relation to building activities undertaken in the traditional mainstream), how might there be a significant step-change of this situation in the future? What could best enable the community-led sector to increase its role?

## Enabling the national community-led housing sector

The vision and objectives that emerged from the collaboration between members of the Community Led Homes alliance[1] was an inheritance from previous recommendations for such a national development, expressed by the Commission for Co-operative and Mutual Housing in 2009.[2] Whilst conforming closely with the broad spirit of that previous national assessment, Alliance members underlined a ready pragmatism for accepting that there would be more than one way for community-led housing projects to come forward. Projects could come through, or be initiated by:

- new start-up community groups;
- supportive local authorities or local property developers (including housing associations);
- existing 'community anchor' bodies (with tangible connections to local communities from providing other community services) seeking to become more engaged with local housing provision.

**Table 5.1: Benchmarks for assessing the impact of community-led housing projects**

| The underlying challenge to ideas for community-led housing (CLH): | Potential benchmarks of a positive impact from CLH activities: |
|---|---|
| Domination of UK housing delivery by mainstream and traditional bodies | Sign-up and promotion of community-led routes by public sector leaders |
| Scepticism on the extent of demand from people to provide their own homes | Inclusion of CLH ideas into 'needs assessments' and other formal surveys |
| Risks routinely perceived to be present with inexperienced projects | Development of local 'enabling' frameworks to develop CLH skill-sets |
| Limited number of formal partnerships working with community initiatives | Awareness-raising community-led presentations to training/CPD seminars |
| Monopolisation of land resources by mainstream interests | The adoption of statutory policies which identify sites for community schemes |
| Views that 'self-build' activity is only for middle- and higher-income households | Housing providers supporting self-build schemes for lower income groups |
| An inflationary impact on the value of the land, sites or property that CLH projects would use | Land and property prices established by adherence to statutory policy |
| Complexity in acquiring the capital funds to carry a project to completion | Investment funds, loans and bond schemes available to help CLH projects |
| Rigid frameworks through which local 'affordable housing' delivery is framed | Local policies accept community proposals as 'affordable' outcomes |
| Limited skills to operate appropriate governance of property schemes | Advice on selecting appropriate legal formats to suit different CLH 'models' |
| Low availability of start-up funds to help a CLH project meet first milestones | 'Start-up' funds created at local levels to assist CLH projects to commence |

**Table 5.1: Benchmarks for assessing the impact of community-led housing projects (continued)**

| The underlying challenge to ideas for community-led housing (CLH): | Potential benchmarks of a positive impact from CLH activities: |
|---|---|
| Minimal understanding of development and construction roles for build projects | Awareness-raising and training modules to CLH groups and local communities |
| Minimal access to construction and building agents | 'Enabling' support for projects to review options: off-site; design & build; and so on |
| A lack of project-management capacities to organise on-site tasks | Skill development programmes for CLH groups and their project managers |
| A lack of management capacities/ expertise to steer long-term provision | Skill development for new-build, tenant-management, or empty homes schemes |

Until recently, the kind of development finance that could flexibly support initial works and partnerships to progress an idea for scheme development was lacking. This became the basis of lengthy negotiation with the Ministry of Housing, Communities and Local Government for resources to help expand the knowledge, skills and practical engagements of local community initiatives across England. Eventually a commitment was extracted from central government representatives that funds could be found for such purposes through an extension of Community Housing Fund finance, which had first been issued in 2016 to selected local authorities known to have experienced housing market pressures from high numbers of 'second'/holiday homes.

In 2018 the terms of use for this new finance were agreed,[3] with a deadline for all expenditure to be completed by March 2020. The largest element of these funds would be directed via Homes England to very local projects and enabling activities – either as *revenue* funds to support the first stages of development of named community projects (principally covering costs of legal incorporations, planning and design work, and other capacity-building activities) or to promote 'enabling' activities by supportive local authorities, or as *capital* funds to be applied to the cost of physical works, like infrastructure requirements to enable problematic sites to be useable, or for the purchase of existing properties.

One assessment of the impact that has already resulted from the application of these funds, as assessed by Homes England[4] in 2019, was a clear identification of the detailed momentum from local projects looking to deliver almost 19,000 new dwellings during the next five-year period.

Alongside the specific sums for local project-related expenditure, however, was finance to expand the national CLH sector's support networks in very practical ways:

- the creation of a central source of advice and information on CLH practice and initiatives (overseen by the UK Cohousing Network for the CLH alliance);
- the commencement of an accredited national training programme to develop the skills of community-led housing 'facilitators' (overseen for the CLH alliance by the Confederation of Co-operative Housing, and recognised for student and CPD purposes by the Chartered Institute of Housing);
- smaller 'start-up' grants to assist completely new groups establish their initial ambitions and consolidate intended roles and partnerships (overseen for the CLH alliance by Locality and the National Community Land Trust Network);
- grants and peer support for the creation of local and sub-regional community-led housing 'hubs' to become local centres of excellence for practical skills and enabling activities (overseen for the CLH alliance by the National Community Land Trust Network).

The significant contrast between the Community Housing Fund and previous programmes of government support for 'community initiatives' would therefore be both the provision of the *revenue* funds in the form of a facilitative grant (not a loan), and the funds provided to help establish the national 'infrastructure' of support and advice services. The one unsatisfactory element in this situation would always be the relatively short timescale for their use in the fiscal period set by HM Treasury.

## Fostering impact from local community-led housing 'hubs'

Notwithstanding the role of the national bodies to allocate CHF funds to local projects, the greatest practical on-the-ground support and impact will come from the growing engagements of the new CLH 'hubs'. It will be crucial that they demonstrate how collaborative partnerships between different levels of support for community-led initiatives can operate with clear lines of accountability back to the local and community interests they support. It is recognised that Hub operations will need to be engaged with key stakeholders, such as statutory authorities, local housing delivery providers and housing associations, but they are still pictured to be *stand-alone* entities, outside of control by any one mainstream or established partner.

For those familiar with previous engagements of community-sector bodies, like the work of 'secondary housing co-operatives' in the 1960s and 1970s, or the community self-build focus of the Walter Segal Trust and the Young Builders Trust in the 1990s, there might be some uncertainty about what new impact these hubs could achieve. A look at the indicative services coming from the new bodies[5] will, however, recognise the intention to provide a comprehensive set of enabling services, where these can be established, including:

- administering 'start-up grants' and other deposits for legal registrations;

- advising on forms of legal incorporations;
- assessing information and property searches on potential sites for CLH-projects;
- undertaking or facilitating property assessments and surveys;
- providing a service as financial intermediaries;
- acting as a 'portal' for the engagement of professional expertise and agents;
- assisting with the formation of project partnerships;
- acting as advocate for a local group or initiative;
- undertaking financial assessments of scheme feasibilities;
- compiling and submitting project applications;
- project management duties through all start-up, design and construction stages;
- skills development for groups to manage built properties and neighbourhoods;
- assisting with local policy development of statutory and other stakeholders.

The effective skills demanded from such hubs could therefore cover a wide range of practical issues, for both inward- and outward-facing responsibilities. This will involve at times being pragmatic in ways to help specific initiatives, and at times to act as an advocate for the community-led concept at large, in all its manifestations, and to challenge doubts that could issue from potentially critical observers.

At a very practical level, community-led housing hubs will also need to address how the economic consequences of community-led schemes are presented within scenarios that are not of their own making. A critique about community-led schemes is that the consequence of the involvement of unskilled activists could easily mean their development costs will turn out to be higher than other development projects – and therefore less viable. Notions of 'viability' in the UK's property development sector have become pronounced in recent years, fuelled not least by the impact of policies from the Conservative and Coalition governments[6] that have inflated the expectations of landowners and developers of the financial return and profit that should be obtained from their engagements. The standards of the UK's private sector development bodies are not, however, appropriate for all projects and it would be one-dimensional to assess the potential returns from community-led projects purely in monetary terms. Ideas of '*social* capital' and 'whole-life' costings are also relevant here, alongside assessments of what it could cost *not* to change local circumstances,[7] and are all legitimate factors for assessing the full costs and returns of community-led schemes. At the very least, a full cost–benefit analysis should compare like with like, and not use an isolated measure or blanket scepticism as an excuse to disregard what community-based projects could achieve.

Nor is it the case that the nature of community-led housing developments will *reduce* the value of residential land or depress what could be gained from a site's residential development. Claims can be encountered from agents within the UK's property development sector that the full monetary value of a site's building development capacity will be the total future receipts stemming from 100% of future residential dwellings being sold at open market rates: any departure from that outcome is then adjudged to be a reduction in the site's true worth and in fact would represent an unmerited subsidy being demanded from the landowner or the construction company towards households or future unknown interests.

This depiction of basic land values is, in part, itself a legacy of the contemporary debate relating to the provision of affordable housing and to the UK's private housebuilding and development sector[8] arguments that housing development projects could be made unviable if they are slavishly required to support local policy requirements, such as supplying a minimum level of affordable units. Negotiating a negation of such requirements has been a means to inflate a site's development value, especially if the development outcomes are to come solely by providing all the properties for sale at open market prices. Yet suggestions that local policy requirements for 'less-than-open-market-priced' units somehow represent a form of *sub*-market delivery are quite misleading – policy sets the requirements for a site's use, and the terms and nature of that use contribute to a site's value, especially when viewed within its market context. Being able to avoid some policy obligations is not a means to realise a site's 'true' economic worth – it is merely an opportunistic means to squeeze more outcome or profit from what was previously intended from a consistent use of policy for all sites or land of that kind.

There is a role here for community-led hubs to support local planning authorities in challenging the kinds of destabilising forces noted here. In accepting that the value of land or property is clearly set within the nature of the planning policy approved for such development, hubs can represent a distinct willingness to build in compliance of that policy and thereby challenge attempts to discredit what the local planning policies are seeking to put into place. The fact that community-led schemes could accept the formal requirements demanded by local planning policies is not by itself grounds to conclude that such schemes will be more costly than other 'open market' development, especially if the comparison is with practices that historically have tried to undertake development 'on the cheap' and bully their way forward to force a non-compliance with core policies.

The hubs will certainly need to show how community-led projects can deliver within the economic parameters of a policy-compliant scheme, and that they can still create viable outcomes regardless of their constituent costs. Where community-led schemes can positively address the economics of their projects is in a readiness to accept a level of overall profit different from that which is routinely sought by the private sector. Hub development will also have an opportunity to benefit from the unpaid contributions made by community or volunteer management, which can assist scheme savings in some instances. None of this makes CLH projects automatically less viable than 'open market' building projects – comparisons between the approaches really need the elements under examination to be fairly compared.

## Notes

1 As introduced in Chapter One.

2 Commission for Co-operative and Mutual Housing (2009), 'Bringing Democracy Home', Commission/CCH, Liverpool, UK.

3 MHCLG (2018), *Community Housing Fund Prospectus,* HMSO, London.

4 'Homes England call for CLH/CHF "pipeline" information', May 2019, www.gov.uk/government/publications/call-for-information-on-community-led-housing-projects-by-may-31st.

5 Duncan, P. & Lavis, J. (2018), 'Delivering a Community-led Housing Enabling Hub Service: Experience and Lessons from Existing Practice', Power to Change Research Institute, London.

6 As detailed in the National Planning Policy Framework in 2012, and in subsequent government circulars.

7 Such as a likely reduction in payments of exorbitant private sector rents being met either by public funds like local housing allowance payments or by increasing and unsustainable levels of personal incomes, or the reduced running costs of high-quality low-impact dwellings.

8 See Archer, T. & Cole, I. (2016), 'Profits before Volume? Major Housebuilders and the Crisis of Housing Supply', Centre for Regional Economic and Social Research, Sheffield Hallam University, UK; Shelter (2017), *Slipping Through the Loophole: How Viability Assessments Are Reducing Affordable Housing Supply in England,* Shelter Publications, London.

# SIX

# Final remarks

The social and political environments in which community-led and self-build projects must operate are complex and challenging. The government's Housing Minister conceded in his address to the UK's National Community-led Housing conference in 2017 that there remain barriers to what local collaborative and community-led projects might achieve, admitting that 'the biggest barriers are almost certainly cultural'.[1] As has been explored in the foregoing pages, fundamental to the UK's house building and planning culture is *who* decides what local people's key needs at the grassroots level may be, and what actions are, or should be, mobilised as a response.

This publication has presented substantial detail on how local housing and neighbourhood projects have challenged the conventional decision-making frameworks that have traditionally assumed a legitimacy to act *on behalf of local communities*, rather than enabling decisions to be undertaken *by* the households resident in their areas. As summarised by the Policy Exchange:[2]

> why is it people don't particularly like the homes that are [usually] built? In one sense it is because they are

> rarely built for people. The architect works for the client and their peers. The developer for the shareholder. The planner for the place. The individuals who eventually live in the homes are a feature of all their thoughts, but a focus of none.

Ironically, the recent central government willingness to lessen fiscal restrictions on local authorities may, however, give greater impetus to mainstream opportunists who are readier to support local authority-controlled development arrangements or new local housing companies[3] than to other community-directed ones. ARCH (the Association of Retained Council Housing)[4] is loudly promoting the profile of an invigorated council housing sector: 'councils are once again leading the charge to help increase supply to reach the Government's ambitions to deliver 300,000 new homes a year'. Senior local authority voices can already be heard questioning whether there will be any need for further community-led initiatives.[5] If there are to be new opportunities for statutory sponsored housing delivery to meet what has been defined as 'local needs', then why bother with alternatives?

The Local Government Association's recent publication[6] to help the statutory sector engage with local housing markets makes no mention of community-led options, or of the community-led models used in tried and tested initiatives, even while the LGA admits that there are a variety of markets and of appetites to which local provision must relate. There is similarly minimal description of 'community-led' initiatives within the guidelines produced by the Princes Foundation[7] for creating large urban extensions.[8] To return to an analogy between the current age and the political reforms of the early 19th century, supporters of the present-day community-led housing sector could be paraphrased as feeling that:

> if we do not state our views about our housing, they say we are indifferent;

> if we build a few dwellings, they say we are insignificant; if we create many dwellings, they say we wish to intimidate them.[9]

Suffice to note here that proposals for further community-led housing initiatives do not stem from a desire to intimidate other established parts of the *status quo*, not least because community-led initiatives are not against new proposals for council housing or ideas for local housing companies or ideas for more housing association development. Nor should they automatically be assessed as a *threat* to the terms on which local authorities engage with grassroots interests. Separate community-led projects do not imply a rejection of new or existing 'council' housing, or of its core values to provide secure and affordable housing, but they do admit to having ambitions that the standard services may not have for themselves. It is to be hoped, therefore, that enthusiasm from statutory bodes towards local community-led proposals will not evaporate simply because new statutory-driven housing action has been given a renewed impetus.

It will be vital that the CLH sector and its new operational hubs can keep alive a vision for what local communities offer to the nation's social and built environments. The aspirations of local people – singularly or collectively, in new-build or in renovation projects, in the building of single properties, or groups of homes, or even of whole neighbourhoods – amply demonstrate the reality of the following point:

> when building new developments, the aim, therefore, should not be a decisive break from the past, but building on the past and reflecting local public will. The difference between a collection of units and a settlement lies in the capacity to meet these needs.[10]

The latest guidance from the Town and Country Planning Association on the implementation of the government's

'Garden City Prospectus' also recognises the worth in engaging local communities: 'one principle is particularly relevant here: strong local vision and engagement – designed and executed with the engagement and involvement of the existing local community, and future residents and businesses'.[11]

The report of the 'Building Better, Building Beautiful Commission' goes even further, describing how local stewardship and influence from local people is crucial to creating proper places in which communities can thrive, noting there is a real need 'to permit a greater range of self-build, custom-build, community land trusts and other market entrants and innovators to act as developers. In this way our planning system will better respond to the preferences of people as a whole'.[12]

The 'preferences of the people' are many and multifaceted. What has been described in the preceding pages and chapters is just what should be expected when there are the genuine opportunities to put them into practice.

## Notes

1. www.gov.uk/government/speeches/community-led-housing.
2. Policy Exchange (2018), 'Building More, Building Beautiful', London, p 22.
3. Morphet, J. & Clifford, B. (2017), 'Local Authority Direct Provision of Housing: Summary of Research Findings', Royal Town Planning Institute, London.
4. ARCH (www.arch-housing.org.uk/about.aspx).
5. Raised at 'East Midlands Community Led Housing' seminar for senior local authority councillors, 12 July 2019.
6. Local Government Association (2019), 'Understanding Local Housing Markets: Advice and Guidance for Local Authorities', LGA, London.
7. Princes Foundation (2019), 'Housing Britain', Princes Foundation, London.
8. It does promote the 'engagement of local inhabitants' as a general principle for area development.
9. Reformulated by Field M. from Brook, M. (1973), 'The Great Reform Act', p 214, Hutchinson, London.
10. Policy Exchange (2018), 'Building More, Building Beautiful', London.

[11] Town and Country Planning Association (2019), 'Practical Guides for Creating Successful New Communities Guide 11: People, Planning and Power', TCPA, London, p 8.

[12] BBBBC (2020), 'Living with Beauty: Promoting Health, Well-being and Sustainable Growth', The Report of the 'Building Better, Building Beautiful Commission', London, p 2.

# Appendix: research into statutory strategies to help collaborative housing projects

A series of desktop research studies[1] conducted during 2014–17 examined a variety of formal public sector 'housing' and 'planning' strategies for any explicit reference to housing delivery by 'self-provided' or 'collective' routes of action. The examination sought to identify reference to 'mutual' or 'collective' or 'self-provided' activities to develop new local homes and/or neighbourhoods – specifically if there was consideration or mention of any of the following models: *self-build; custom-build; self-help; co-operative; community land trust; cohousing; community-led housing.*

A review of local documentation within Greater London, Greater Manchester and the core of the West Midlands (here termed Greater Birmingham for simplicity's sake)[2] looked through all the relevant documents that were available on the websites of the 49 local authorities within the three metropolitan areas. At least one document from each of the following three categories was open to examination from each authority:

- documents denoting local 'planning' principles – variously termed core strategies, core spatial strategies, local development frameworks, and local plans;

- documents denoting key 'housing' principles – variously termed housing strategies, housing needs assessments, strategic housing market assessments, and affordable housing policies;
- documents mapping out priority engagements with 'local communities' – variously termed community strategies, sustainable community strategies, community plans, corporate plans, voluntary sector strategies, and engagement and involvement strategies.

The overall results of this examination of policies across the three metropolitan areas are summarised at the end of this piece. From all the documents examined, almost three-quarters of the authorities reviewed made no mention of any of the activities or 'models' of community engagement under consideration. Only 13 authorities (about 27%) referred to a self-provided or 'community-led' activity for housing and neighbourhood change. Of these, the predominant mention was for 'self-build' activity, which had double the number of the next highest mentions, namely for 'custom-build' activity. A very small number of authorities had a mention of two separate kinds of activity – linking mention of 'self-build' and 'custom-build' housing together. There was no specific mention at all of models or activities like 'self-help housing' or 'cohousing'.

A 2017 review of strategies in the rural areas of Leicestershire and North Yorkshire[3] (including for any 'Right to Build' strategies) reinforced there being only a rudimentary recognition in statutory policies of the practices engaged in by local communities at large. Besides the self-build mentions, there was a brief reference to 'community land trusts' and to 'cohousing', but there was little recognition of what practicalities would help households and communities to progress specific kinds of projects.

What was not evident in any authority document was an explicit 'people-first' policy position in support of local communities progressing flexible local housing solutions of their own choosing, in response to primary aspirations exhibited at the grassroots level.

**Table A.1: The identification of community-based housing within public sector policies**

| Greater Birmingham Metropolitan area (six authorities examined) | | | | | | | |
|---|---|---|---|---|---|---|---|
| | Self-build | Custom-build | Self-help | Housing co-ops | Community-led | Land trusts | Cohousing |
| Mentions by Birmingham authorities | 2 | 1 | 0 | 0 | 0 | 0 | 0 |
| Greater London Metropolitan area (33 authorities examined) | | | | | | | |
| | Self-build | Custom-build | Self-help | Housing co-ops | Community-led | Land trusts | Cohousing |
| Mentions by London authorities | 4 | 2 | 0 | 0 | 1 | 3 | 0 |
| Greater Manchester Metropolitan area (10 authorities examined) | | | | | | | |
| | Self-build | Custom-build | Self-help | Housing co-ops | Community-led | Land trusts | Cohousing |
| Mentions by Manchester authorities | 3 | 1 | 0 | 1 | 0 | 0 | 0 |
| Summary of mentions from all areas | | | | | | | |
| | Self-build | Custom-build | Self-help | Housing co-ops | Community-led | Land trusts | Cohousing |
| Size of sample total – 49 | 9 | 4 | 0 | 1 | 1 | 3 | 0 |

## Notes

1. The material in this Appendix reflects some of the work issued as the chapter by the author entitled 'Models of Self-Build and Collaborative Housing in the UK', in Benson, M. & Hamiduddin, I. (eds), *Self-Build Homes: Social Discourse, Experiences and Directions*, UCL Press, London.
2. Field, M. (2015), 'Community-led Bulletin', University of Northampton, UK, provided a list of all documents examined, approximately 190 in total.
3. Subsequent work between 2017 and 2019 by North Yorkshire local authorities and their partners has been more expressive of positive planning in support of community-led housing projects.

# Index

## C

## I



## K



## L

## T

Printed and bound by CPI Group (UK) Ltd, Croydon, CR0 4YY

27/11/2023

08195790-0001